The twentieth century has seen biology come of age as a conceptual and quantitative science. Major functional phenomena rather than catalogues of animals and plants comprise the core of MODERN BIOLOGY; such heretofore seemingly unrelated fields as cytology, biochemistry, and genetics are now being unified into a common framework at the molecular level.

The purpose of this Series is to introduce the beginning student in college biology—as well as the gifted high school student and all interested readers—both to the concepts unifying the fields of biology and to the diversity of facts that give the entire field its unique texture. Each book in the Series is an introduction to one of the major foundation stones in the mosaic. Taken together, they provide an integration of the general and the comparative, the cellular and the organismic, the animal and the plant, the structural and the functional—in sum, a solid overview of the dynamic science that is MODERN BIOLOGY.

MODERN BIOLOGY SERIES

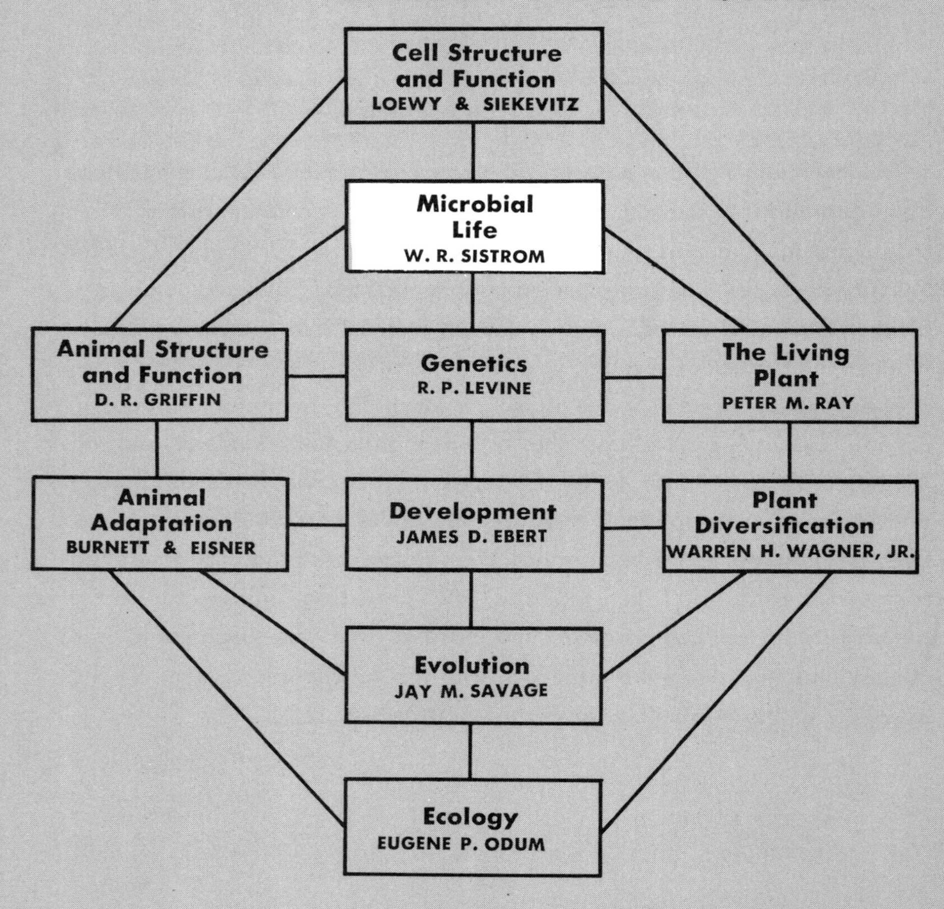

HOLT, RINEHART AND WINSTON

NEW YORK - CHICAGO - SAN FRANCISCO - TORONTO - LONDON

MICROBIAL LIFE

W. R. SISTROM

HARVARD UNIVERSITY

Library of Congress Catalog Card Number: 62-16934

27940-0112
Printed in the United States of America

Cover photograph, *Globigerina,* courtesy of
the American Museum of Natural History

PREFACE

"Ruder heads stand amazed at prodigious and Colossean pieces of Nature, but in these narrow Engines there is a more curious Mathematicks."

Thus wrote Sir Thomas Browne in 1663, not long after the discovery of microorganisms by Anton van Leeuwenhoek. His judgment has been sustained by history. This little book is an attempt to explain the workings of the narrow engines.

In writing the book I have had one major purpose in mind: to make the reader aware of microorganisms, especially of bacteria, as living and growing creatures. To this end, some aspects of bacterial physiology have received greater emphasis than is perhaps usual. This is particularly true of Chapter 6, which deals with growth and protein synthesis. Since making protein and growing are really what a bacterium does, this emphasis is not unwarranted.

I have assumed that the reader is familiar with material in the area of genetics and the cell. *Genetics* and *Cell Structure and Function* are forthcoming books in this Series that will provide such background.

Of necessity, much material has been neglected in this book. Fortunately, it is not difficult to recommend further reading on the subject. The following two books will provide the interested student with more detailed treatments of the topics treated briefly or not covered here: *The Microbial World,* Stanier, Doudoroff, and Adelberg (Prentice-Hall, 1957), and *The Microbes' Contribution to Biology,* Kluyver and van Niel (Harvard University Press, 1956).

A word of appreciation is due Helen E. Speiden for her preparation of all the line drawings.

W. R. S.

Cambridge, Massachusetts
June, 1962

CONTENTS

MICROBIAL
LIFE

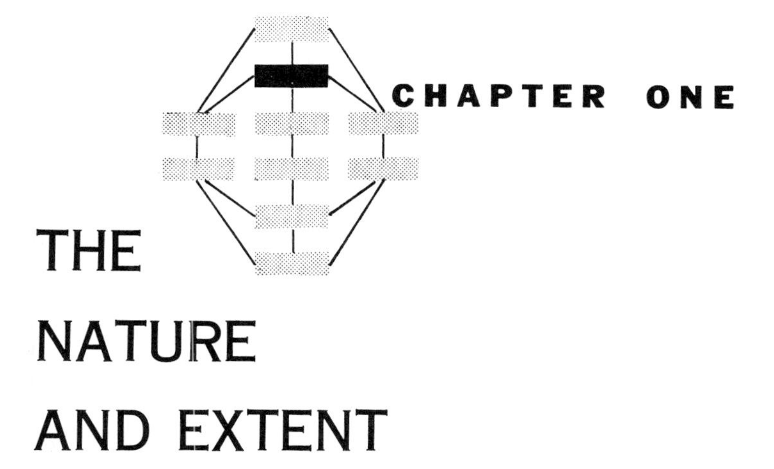

CHAPTER ONE

THE NATURE AND EXTENT OF THE MICROBIAL WORLD

What is a microorganism? There is no simple answer to this question. The word "microorganism" is not the name of a group of related organisms, as are the words "plants" or "invertebrates" or "frogs." The use of the word does, however, indicate that there is something special about *small* organisms; we use no special word to denote large animals or medium-sized ones.

Before we consider what the importance of being small is, we must emphasize that we are talking about small *organisms.* There are many objects of biological interest that are small: the cells of plants and animals are the most obvious examples. However, these are not organisms but only parts of organisms. A microorganism is small, but it is also an individual.

Although it is measured in terms of length and breadth and volume, the size of an organism is significant only in terms of chemistry and metabolism. The most obvious result of being small is a very large ratio of surface area to volume. For example, a volume of bacteria equal to the volume of a single plant cell has a surface area roughly ten times as great. This is important because it is through its surface that an organism exchanges material with its environment. For a given mass of cells a larger surface means a more rapid exchange and a more intense metabolism within the organism. Although other factors probably play a role, the extremely intense metabolism and high rates of reproduction characteristic of microorganisms stem primarily from the high ratio of surface area to volume that is inherent in their small size.

PLANTS, ANIMALS, AND PROTISTS

We have said that the term microorganism does not define any particular group of living things but is merely descriptive of certain kinds of organisms. What is the relationship of microorganisms to the animals and plants of which you are most aware? The familiar division of living things into plants and animals is probably as old as human observation; however, this is no guarantee of its validity. Biologists have long accepted this common-sense division and formalized it by creating the two kingdoms of Plants and Animals. Since this is a common-sense division, any reasonably observant person can quickly set down several criteria that distinguish plants from animals. For example: animals move, whereas plants do not; plants are green and do not eat other organisms, whereas animals do eat plants or other animals and are seldom green; plants usually have no fixed size, whereas animals do; and so forth.

This classification was made before any real information about microorganisms had been gathered. Although microorganisms had been observed by Anton van Leeuwenhoek as early as 1674, it was not until the development of the compound microscope in the eighteenth and especially in the nineteenth centuries that biologists became aware of the tremendous numbers and diversity of organisms whose very existence could be demonstrated only with the microscope. As information accumulated it became clear that the old and familiar division of the living world into plants and animals would not suffice. Where, for example, would one put a creature which was green and did not eat other organisms (so far, clearly a plant) but which was highly motile (thus animallike); or again, a creature which did not move and yet ate other little organisms; or yet again motile organisms which were neither green (thus not plants) nor ate other creatures (and thus not animals either). It should be emphasized that these difficulties, although acute with microorganisms, are not peculiar to them. Many large organisms, some fungi for example, have as anomolous a position.

As early as 1894 the German biologist Ernst Haeckel suggested that a way out of this impasse would be to create a third kingdom, to encompass those organisms which were not obviously either plants or animals. He proposed the name Protista for this kingdom. At that time the only criterion for placing an organism in Protista was the negative one that it could not easily be put in either of the other two. As our knowledge of these organisms advanced, it became more and more clear that two kingdoms are inadequate; and also further criteria for the third kingdom became apparent.

The Protista includes all organisms that do not have any extensive development of tissues, that is, systems of distinctive cell types performing different functions. This kingdom includes, then, all unicellular and microscopic organisms; but it also includes many multicellular organisms and some very large ones. Although devoid of tissue systems, multicellular protists are not necessarily simple in structure. For example, some of the large seaweeds (algae) are composed of a long, thin stem or stipe, to which are attached broad leaflike blades called fronds, but the cells within the stipe and within the fronds are basically very similar in structure. Like most classifications in biology, this one is not absolute. In the example just mentioned there is obviously a tendency toward the development of tissue systems; however, this is not fundamental to the structure of any of the protists. The Protista includes the algae, the fungi, the protozoans, and the bacteria.

The protists themselves can be divided into two large groups, which for simplicity can be termed the higher and the lower protists. The lower protists include the bacteria and blue-green algae. This division is based on fundamental and far-reaching differences in cellular organization which will be considered in some detail in Chapter 3. The most important of these differences has to do with the structure of the cell nucleus. In the cells of all organisms, except the bacteria and blue-green algae, the nucleus is surrounded for most of the life of the cell by a nuclear membrane. This membrane is not found in the lower protists; in these organisms the nucleus is simply embedded in the cytoplasm. The bacteria and the blue-green algae are all microscopic.

The broad outlines of the Protista and their relationships with plants and animals are shown in Fig. 1-1. The various groups in the Protista will be treated in more detail in Chapter 2. It will be seen that many of these groups include organisms which are not microscopic. It is clear, then, that the term "microorganism" is not a very useful one. The place of microorganisms in the living world can be understood only if one understands the place of the protists of which they are a part.

PRODUCERS, CONSUMERS, AND MIDDLEMEN

The common-sense division of plants and animals is based on what appear to be largely morphological or structural characteristics: motility, color, mode of growth, and so forth. These structural differences, however, reflect differences in the ways of living of plants and animals. Plants, of course, are photosynthetic; that is, they utilize energy of sunlight to convert carbon dioxide and water into cellular material and oxygen. Plants obtain the energy needed for growth from sunlight and their carbon from carbon dioxide. This "fixation" of radiant energy and carbon dioxide is absolutely indispensable to the continued existence of all life on earth. Plants are *producers;* they produce "fixed" energy and carbon in the forms of their own substance, and also oxygen that *consumers* can then use. Animals are consumers; they live by eating either plants or other animals. The carbon of a consumer comes directly

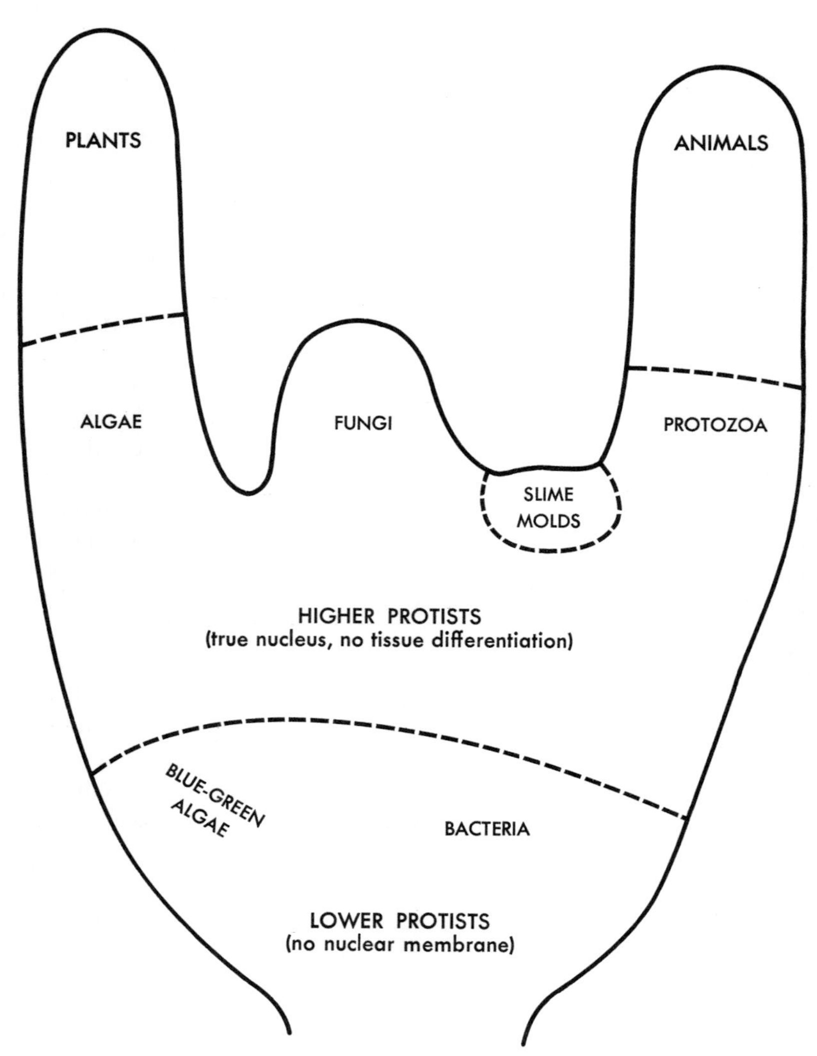

Fig. 1-1. The place of Protistà in the living world.

or indirectly from the carbon in a plant product; the energy used by a consumer comes from the oxidation or burning with oxygen of some of the plant material back to carbon dioxide and water. Most of the criteria commonly used to separate plants and animals are reflections of these two modes of life. For example, plants have no need to move, since both sunlight and carbon dioxide are pretty uniformly distributed over the surface of the earth. To eat other animals or even plants, animals clearly must be able to move.

The protists include both producers and consumers. The algae are perhaps the most important producers of all; the protozoans are typical consumers, pursuing and eating algae or other (and smaller) protozoans. Among the protists are organisms that are not producers yet differ importantly from consumers. As we shall see, these organisms are indispensable intermediaries between the producers and consumers and are aptly termed "middlemen."

We have seen how animals oxidize a part of the plant material they consume to obtain energy, thus returning some of the carbon to the atmosphere as carbon dioxide. However, only a fraction of the carbon is thus directly recycled, most of it becomes immobilized in the vast array of substances that make up the bodies of plants and animals. When the organisms die, this carbon must be returned to the atmosphere as carbon dioxide so that the cycle may continue and life go on. In other words, between the consumers and the producers must come the middlemen. This role is played by two groups of protists: bacteria and molds.

The bacteria and molds display as a group and as individuals a remarkable catholicity of taste. There is literally no known organic constituent of living things that cannot be used as food by at least one kind of bacterium or mold. This is clearly essential to the continuing flow of carbon; otherwise carbon and energy would slowly but certainly accumulate in unused products of plant or animal metabolism. Just as many of the distinctive characteristics of plants and animals are reflections of their roles as producers and consumers, so also are the characteristics of molds and especially of bacteria reflections of their role as middlemen. As will be shown in subsequent chapters, the metabolism of bacteria is marked by an enormous complexity and diversity. It will also be shown that underlying this complexity is a fundamental oneness of mechanism shared by all living creatures. Another unity in bacterial metabolism comes from the important role of bacteria in nature as the indispensable feeders of the furnace of photosynthesis.

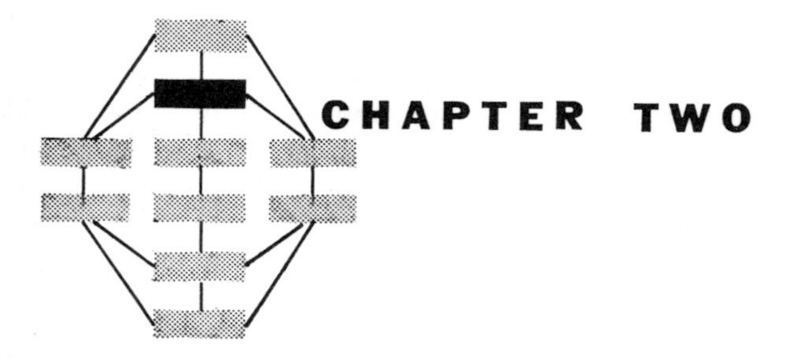

CHAPTER TWO

A SURVEY OF THE PROTISTA

In the first chapter the kingdom Protista was defined and its relationships with other major groups of organisms were examined. Let us now look in a little detail at the various groups of protists shown in Fig. 1-1. Unfortunately we can do no more than select a few examples from each major group.

THE HIGHER PROTISTS

The higher protists include fungi, protozoans, and algae (excepting the blue-green algae). Remember that these organisms are distinguished from the lower protists (bacteria and blue-green algae) by having a nucleus which is surrounded by a membrane.

The Fungi

Such common and well-known creatures as mushrooms, yeasts, and molds are fungi. As mentioned earlier the fungi play a role similar to that of the bacteria in the economy of nature: they are typically organisms of decay. Unlike the algae they do not photosynthesize, and unlike most protozoa they utilize soluble foodstuffs only.

A fungus is composed of innumerable long, fine filaments of cells called hyphae which together constitute the mycelium. The hyphae may be tightly interwoven, producing a tough, compact mycelium. Most fungi are nonmotile, although a few produce motile reproductive cells. Sexual reproduction is common in this group and, in addition, all fungi have some mode of asexual reproduction. There are three main groups of fungi: the phycomycetes, the ascomycetes, and the basidiomycetes.

The phycomycetes include both aquatic and terrestrial organisms; the other fungi are exclusively terrestrial. Aquatic phycomycetes (or water molds) possess motile reproductive stages, whereas the terrestrial forms, in common with the rest of the fungi, do not possess any motile stages. Some of the simplest of the water molds do not have the typical mycelial habit of the fungi, but are merely saclike bodies with a few short filaments (rhizoids) that penetrate the substratum. The hyphae of the mycelial phycomycetes are not regularly divided by cross-walls; these are formed only to delimit reproductive organs. Thus the mycelium contains many nuclei in a common cytoplasm. The mycelia of the other two groups of fungi—the ascomycetes and the basidiomycetes—are septate, that is, regularly divided by crosswalls.

The ascomycetes and basidiomycetes are distinguished from each other by the nature of the sexually produced fruiting body. Figs. 2-1 and 2-2 show

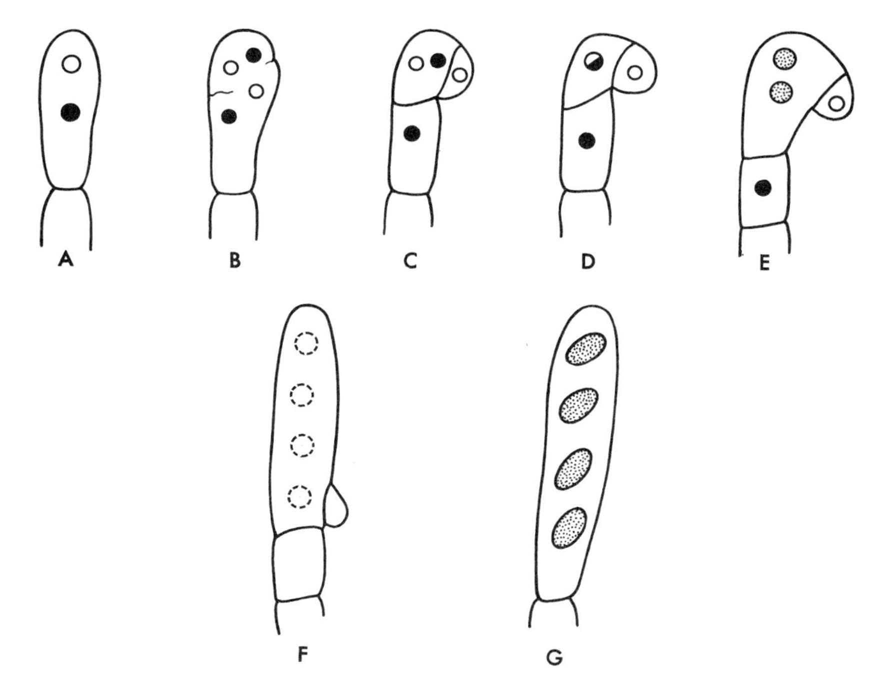

Fig. 2-1. The formation of ascospores. A: A cell at the tip of a hypha containing two nuclei, one from each parent. The origins of the nuclei are indicated by shading. B and C: Each nucleus divides and three cells are formed, two with one nucleus each and the third with two nuclei, one from each parent. The ascus will develop from this cell. D: The two nuclei fuse to produce the zygote nucleus ($2n$), which immediately (E and F) undergoes meiosis. The resulting four haploid ($1n$) nuclei are included in the four ascospores finally produced (G). In many ascomycetes the four nuclei each divide once or twice more (mitotically) giving rise to 8 or 16 ascospores.

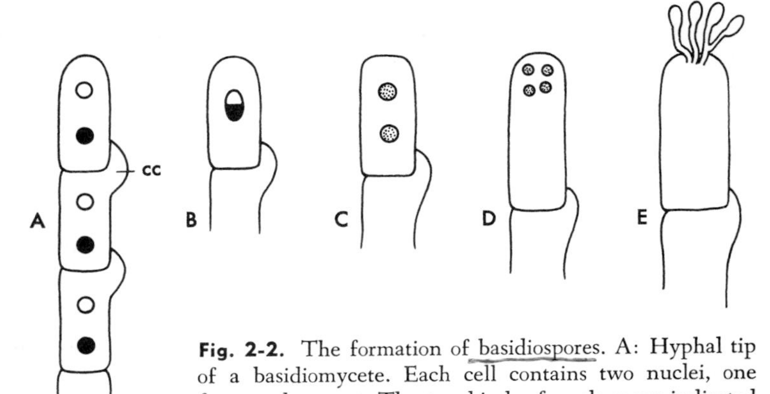

Fig. 2-2. The formation of basidiospores. A: Hyphal tip of a basidiomycete. Each cell contains two nuclei, one from each parent. The two kinds of nucleus are indicated by the shading. B: The two nuclei of the terminal cell have fused to produce the zygote nucleus ($2n$), which immediately divides meiotically (C and D) to give four haploid nuclei ($1n$). These nuclei are included in the four external basidiospores finally formed (E). Label *cc* indicates the clamp connections formed between cells to insure that each newly formed cell receives a nucleus from each parent.

diagrammatically the stages in the formation of the two kinds of fruiting bodies, the ascus and the basidium. The principle difference is that the ascospores are contained within the ascus, while the basidiospores are borne externally on the basidium. In many ascomycetes special sex cells or gametes are produced in morphologically distinct structures, the gametangia. In other ascomycetes and in most of the basidiomycetes there are no specialized sex cells; any part of the mycelium can take part in sexual reproduction. The life cycles of these fungi will be discussed in Chapter 7.

The basidiomycetes are further distinguished by a peculiarity in their sexual reproduction. In all other organisms the *haploid* nuclei of the male and female gametes fuse immediately after fusion of the gametes themselves to form the *diploid* zygote nucleus. In the basidiomycetes nuclear *fusion* is delayed; however, the two kinds of nuclei (one from one parent, the other from the other parent) continue to divide conjointly. This results in a unique kind of organism called a *dikaryon.* The individual nuclei of a dikaryon are haploid, but the organism as a whole contains the complete set of genes from both parents. A special morphological feature is present in the dikaryon: the clamp connection (Fig. 2-2). This structure provides a way for the four daughter nuclei formed during cell division to assort themselves so that each daughter cell receives one of each kind of nucleus.

The Algae

The algae are the most important "producers" of aquatic environments. The ability to photosynthesize is the fundamental physiological characteristic of the algae, and consequently all algae are colored. Except in the blue-green algae, the photosynthetic pigments and the essential reactions of photosynthesis occur in subcellular organelles called chloroplasts, similar to those of higher plants (Fig. 3-1). There are two main kinds of chloroplast pigments: chlorophylls and carotenoids. All algae contain chlorophyll *a* and, in addition, one or more other chlorophyll pigments. These pigments are green. Various carotenoid pigments, which are yellow or red, are also found. The carotenoid pigments impart distinctive colors to algae. A third kind of chloroplast pigment, the phycobilins, occurs only in two groups, the blue-green and the red algae. The algae may be classified into several large groups on the basis of color. Thus, for example, there are the green, yellow, brown, and red algae. These colors reflect different patterns of chloroplast pigments. Usually, the nature of the food-reserve material produced in photosynthesis is characteristic in each group.

Sexual reproduction is common in the algae. In unicellular forms there are no specialized sex cells; in the multicellular algae special sex cells, which are usually motile, are formed. Many multicellular algae can reproduce asexually by the production of motile cells called zoospores.

In the great majority of cases, motile cells of algae possess flagella. These are long, whiplike appendages which act somewhat like oars in propelling the cell. In each major group the number and arrangement of flagella is characteristic. The *diatoms* are a specialized group of algae in which motility has an entirely different basis. The diatom is enclosed within a rigid wall that is pierced by a longitudinal slot. The cytoplasm can protrude through this wall and push against the surface.

Apart from these features, the algae are a very heterogeneous assemblage. In size alone, they range from microscopic, unicellular forms to giant kelps, found in the Pacific Ocean, which may be 150 feet or more in length. Because of their variety, we cannot discuss all the different groups of algae, but can only describe a representative of one group. The structure of the unicellular alga shown in Fig. 2-3 is typical of the cells of even multicellular algae. As mentioned in Chapter 1, the cells of large algae are similar to unicellular algae, as shown in Fig. 2-3. *Chlamydomonas* is a typical unicellular, green alga. The cell is about 10 μ (microns) in diameter and is surrounded by a thick cell wall composed of cellulose. The presence of cellulose in the cell wall is characteristic of the green algae; in other algae the wall, if present, is made of some other material. Just within the wall is the cell membrane. The most striking feature of the cell is the large, green, cup-shaped chloroplast. The number and shapes of its chloroplasts are characteristic of each alga. The chloroplast is surrounded by a membrane. The other organism in Fig. 2-3 is the multicellular green alga, *Ulva*.

It was stated in the foregoing section that the fundamental property of the algae is the ability to photosynthesize. Nevertheless, there are several unicellular organisms, best described as colorless algae, that are morpho-

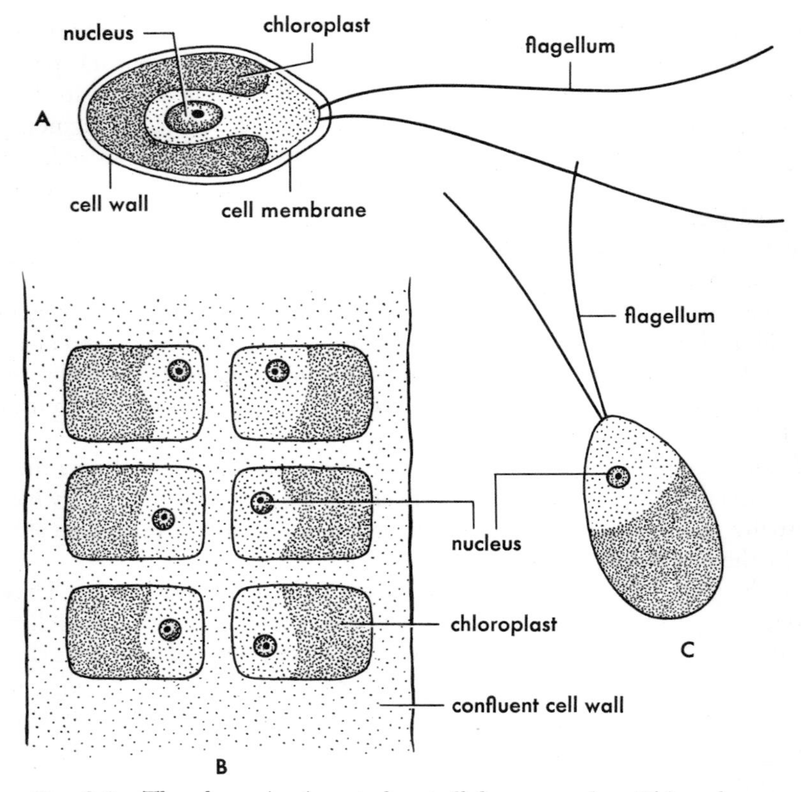

Fig. 2-3. The algae. A: A typical unicellular green alga, *Chlamydomonas*. In B is shown a portion of a multicellular green alga, *Ulva*. This alga consists of a sheet two cells thick that may reach a length of one foot. In C is shown a motile sex cell (gamete) of *Ulva*. Note the similarity of all three cell types.

logically identical to known algae except for the absence of pigments. Some algae when cultivated in the laboratory can be converted into colorless forms that are indistinguishable from some naturally occurring nonphotosynthetic organisms. The arbitrariness of the dividing lines between the various groups of protists thus becomes evident. More important than the arbitrariness itself, however, is the biological reason for it. It seems probable that these pairs of photosynthetic and nonphotosynthetic organisms reflect the fact that both algae and protozoa evolved from a common photosynthetic ancestor.

The Protozoa

This is an enormously complex and heterogeneous group of microorganisms. We can consider here only two representatives: the primitive *Amoeba* and the highly specialized *Paramecium*.

An amoeba does not have a rigid cell wall; the cell is bounded by a flexible pellicle (Fig. 2-4). An amoeba moves by extending a small part of itself forward and then pulling the rest of the cell up into the extension. These extensions, called pseudopodia, also serve for the ingestion of food. A particle of food—for example, a bacterium—is surrounded by a pseudopodium; the particle passes into the cell in a little sac formed of the cell membrane. The food is digested within this sac as it is propelled through the cell by the streaming of the protoplasm. Any portion of the cell surface can form pseudopodia, either for motility or for ingestion of food. Some amoebae have small clear bubbles or vacuoles in the cytoplasm. These slowly fill with a watery solution, move toward the edge of the cell and discharge their contents. These contractile vacuoles are a primitive kind of excretory device. Notice that no one part of the amoeba is specialized for performing these functions.

The simple organization of an amoeba is in marked contrast to the complex cell of the paramecium, a representative ciliate. (Fig. 2-5). The surface of the cell is covered by innumerable short flagella, or cilia. The cilia propel the cell by a coordinated motion similar to that of the oars of a Roman galley. This high degree of integration is accomplished by a rudimentary nerve network consisting of a system of fine filaments that connect the bases of the cilia. Along one side of the cell lies the oral groove, which is also lined with cilia and which leads into the gullet. Food is ingested by being swept along the oral groove and into the gullet, whence it passes into the cytoplasm. The ingested food particle is enclosed in a food vacuole within which it is digested. Remaining indigestible food is excreted through a posterior anal pore. Contractile vacuoles similar to those of an amoeba are also present.

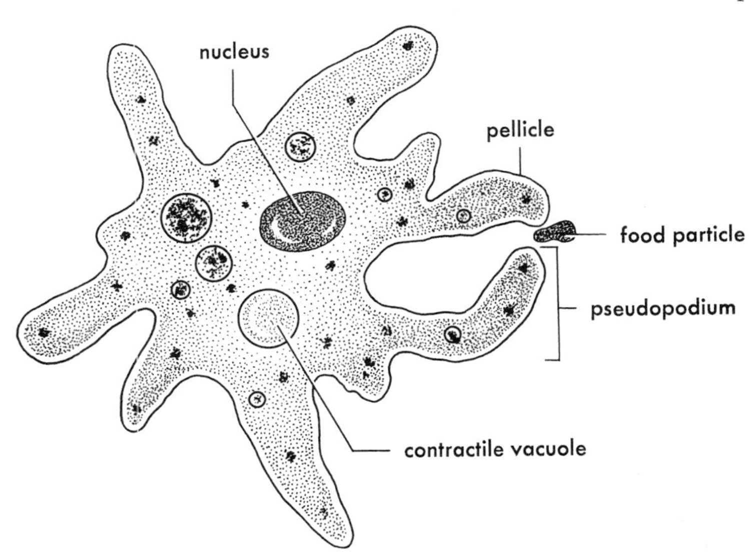

Fig. 2-4. An amoeba, a simple protozoan.

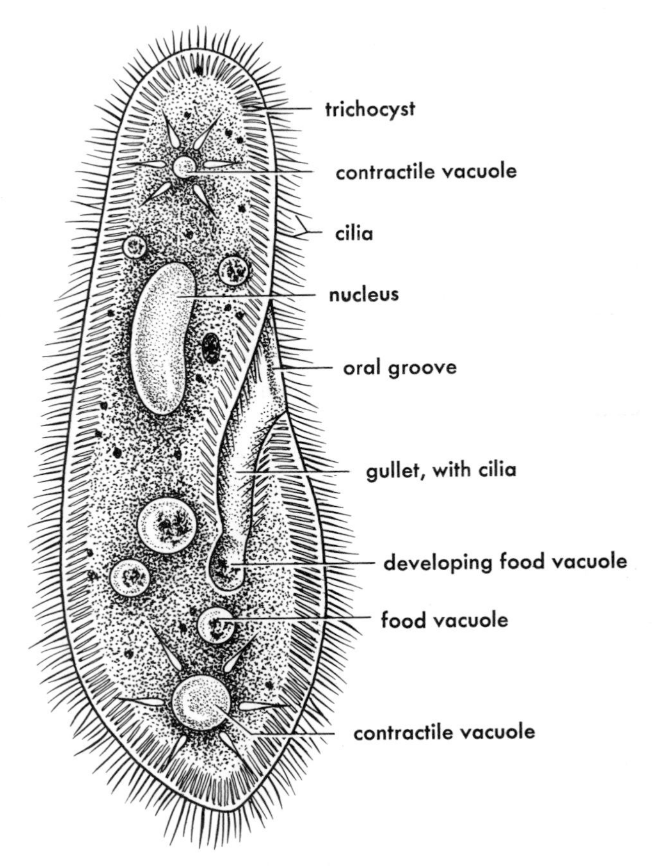

Fig. 2-5. A paramecium, a highly developed protozoan.

From even this brief outline of the structure and behavior of a paramecium it is clear how very different it is from the simple amoeba. The paramecium is, indeed, just about as different from an amoeba as it can be short of becoming truly multicellular. There are limits, apparently reached in the paramecium, to the number of specialized functions which can be carried out by differentiated parts of a single cell. To reach a higher level of organization it is necessary to extend the differentiation to entire cells, that is, to have different functions performed by special cell types. It is this formation of permanently differentiated cell types within a single organism that sets off the higher plants and animals from the protists.

The Slime Molds

The slime molds are a group of organisms of very uncertain relationships. Their outstanding characteristic is a unique life cycle, that involves the formation of a complex and highly organized structure by the cooperative action of a large number of individual cells.

The most conspicuous stage in the life cycle of a slime mold is the fruiting body, which usually consists of a slender stalk with a cluster of spores at the apex. The spores germinate when moistened and a single cell, called a *myxamoeba*, emerges from each spore. The myxamoeba is, in all respects, very similar to a typical protozoan amoeba. It lacks flagella but moves by means of pseudopodia, which also serve to ingest solid food, usually bacteria. The myxamoebae divide by simple fission and soon cover the surface of the substrate. After a time and for reasons that are not yet entirely clear, a few of these cells begin to attract others towards themselves. Soon, thousands of myxamoebae are streaming toward a common point where they form the fruiting body. The individual myxamoebae cooperate in the production of this complex structure; some form the stalk while others are transformed into spores after creeping up the stalk. In a few species the myxamoebae, as they arrive at the center of the aggregation point, organize themselves into a slug-like body that moves as a unit over the surface before finally forming the fruiting body.

THE LOWER PROTISTS

The lower protists are set off from the higher protists by the lack of a demonstrable nuclear membrane. This chapter outlines the salient features of the major groups of lower protists. Our knowledge of many of these forms is fragmentary, and further investigation may reveal that some should more properly be included in the higher protists.

The Blue-Green Algae

The blue-green algae are photosynthetic organisms. The characteristic blue-green color comes from the presence of a blue phycobilin. These algae are all small; some are no larger than a fair-sized bacterium (2 by 4 μ), but most are somewhat larger. They do not have rigid cell walls. The cells often form chains or filaments. The blue-grèen algae are motile but do not have flagella. They can move only on a moist surface; however, the mechanism of this "gliding motility" is unknown. Reproduction is generally by binary fission, but a few of the filamentous types reproduce by a fragmentation of the filament.

As do the higher algae, the blue-green algae have colorless counterparts. Among these may be mentioned *Beggiatoa* which is very similar morphologically to *Oscillatoria*, a typical filamentous blue-green alga.

The Myxobacteria

These organisms, like the blue-green algae, lack rigid cell walls and have a gliding motility but are nonphotosynthetic. The distinguishing feature of the myxobacteria is a peculiar life cycle that has no real counterpart, although it superficially resembles that of some slime molds (see p. 12). The individual cells, which are often cigar-shaped, glide about on the surface of an appropriate medium covering the surface with an almost invisible film of growth. After a while, and for unknown reasons, the vegetative cells begin to aggregate toward a number of centers. As the cells approach each other they heap up, producing simple fingerlike fruiting bodies or, in some forms, very complex treelike ones. Some of the vegetative cells form the stalk, whereas others are transformed into resting forms called *microcysts*. These cysts, when transferred to a fresh medium, germinate, each producing a single vegetative cell, which grows and divides by binary fission to complete the cycle.

The Spirochaetes

These organisms, of uncertain affinities, are also devoid of rigid cell walls. The spirochaetes, unlike the blue-green algae and the myxobactria, are flagellated. They are characteristically helical in shape; the cells divide by binary fission.

The Bacteria

The bacteria include organisms that possess rigid cell walls and when motile have flagella. Three major groups of bacteria can be distinguished morphologically: the true bacteria, the actinomycetes, and the budding bacteria. The true bacteria are discussed in the next chapter.

The actinomycetes

This is a rather large group of organisms, probably closely related to the eubacteria. Actinomycetes are characterized by a tendency toward branching. In some forms the branching is very rudimentary, the cells often being shaped like a Y. In other representatives of this group, extensive branching occurs and moldlike mycelium is formed. However, this is easily distinguished from a fungal mycelium by its very much smaller diameter that does not exceed that of a typical eubacterium. The actinomycetes with extensive branching are further distinguished from the typical bacteria by the formation of a special kind of reproductive cell. These cells, called conidia, are formed singly or in chains at the ends of the branches. They are somewhat resistant to deleterious influences although not to the same degree as bacterial endospores. The actinomycetes are without exception nonmotile.

The budding bacteria

This is a small group of organisms characterized by a unique mode of cell multiplication. The cells are shaped rather like fat footballs, at one end of which is formed a small tube that gradually increases in length. After a while the tip of the tube begins to expand and a new cell is formed. An extensive network of cells connected by these tubes is thus formed.

The Rickettsia

There are many groups of little-known organisms whose relationships are not at all clear. Only one of these, the *Rickettsia,* will be mentioned here. These are very tiny (0.2 to 0.3 μ) obligate intracellular parasites, which are responsible for a variety of serious diseases including typhus and Rocky Mountain spotted fever.

THE TECHNIQUE OF MICROBIOLOGY

The minuteness of the bacteria, and of many other protists, has been stressed repeatedly. You may well wonder how such tiny organisms are handled. How can one study organisms that can be seen only under a microscope? Such minute cells can hardly provide much morphological variety. How can one be sure whether two rod-shaped bacteria are the same or different organisms? The answers to these questions are essential to an understanding of microbiology; to answer them, we must examine briefly some of the basic methodological principles of microbiology. These methods were, to a large extent, developed to handle bacteria, because here the problem of distinguishing morphologically well-nigh indistinguishable forms is most acute. However, these same methods have subsequently been used with other microbes.

The problem is, simply, how can a pure culture be obtained? By "a pure culture" is meant a collection of cells all of which are known to be of the same kind of bacteria. Since bacteria divide by simple binary fission, a pure culture can be obtained by isolating a single cell; the progeny of a single cell will be a pure culture by definition. Note that a pure culture is not defined as one in which all individual cells have identical properties. In the first place, it is difficult to examine the properties of individual bacteria, thus such a criterion could not be met. More importantly, as will be shown later, the properties of the progeny of a single cell are not necessarily identical. A pure culture, then, is defined by the *operation* of isolating a single cell and allowing it to multiply.

How though can a single cell measuring perhaps 1 or 2 μ in diameter be isolated? Recently it has become possible in some instances to do this under the microscope. By means of an instrument called a micromanipulator, a single cell can be mechanically isolated. This method is too cumbersome and involved for ordinary work. More importantly, when one is attempting to isolate a single type of bacterium from a mixed population, the micromanipulator is useless unless the desired form is morphologically distinct.

It is a common observation that molds grow as isolated mycelia on solid substrates, for example, the little flecks of blue mold on cheese or bread. On other solid substrates, such as the cut surfaces of potatoes, a similar development of bacteria is observed. There will be small masses (colonies) of bacteria on the surface. There may be different kinds of colonies, but the bacteria within any one colony are all the same. This observation led to the use of artificially solidified media; Robert Koch was primarily responsible for the development of this method. Solid medium is prepared by introducing a suitable gelling agent into appropriate liquid media; gelatin can be used but more often a substance called agar is employed. Agar must be heated to 100° C before it melts; it solidifies at about 45° C. The still-molten medium is poured into Petri dishes (round glass dishes with a loosely fitting cover) and allowed to cool and solidify. It is then inoculated with a suspension of bacteria by dipping a needle into the suspension then drawing this needle right across the surface of the medium. The bacteria will be wiped off the needle and, at some point, the cells will be separated by a considerable distance. After incubation, colonies will appear on the surface. At the place where the needle first touched the agar the growth will be confluent, but farther out on the streak the density of the colonies will gradually decrease and isolated single colonies will be found. The principle is simply that bacteria, even the motile ones, do not move across the dry surface of the agar. The isolated colonies can only be *assumed* to have arisen from a single cell (two cells in close proximity would give rise to a single colony). Therefore to assure a pure culture, the procedure is repeated using a suspension made from one of the isolated colonies. There are, of course, many variations to this basic technique that need not be elaborated. A pure culture obtained in this way can be transferred to a test tube containing the agar medium.

The isolation of pure cultures rests on the possibility of obtaining media and apparatus that are sterile; in other words, in a condition such that in the absence of inoculation (deliberate or accidental) growth will not occur. The work of Louis Pasteur on spontaneous generation provided the technique and, at the same time, the indispensable demonstration that sterility could be obtained. Entirely apart from the great importance of this work to biology as a whole, this demonstration by Pasteur of the nonoccurrence of spontaneous generation is the fundamental methodological basis for microbiology.

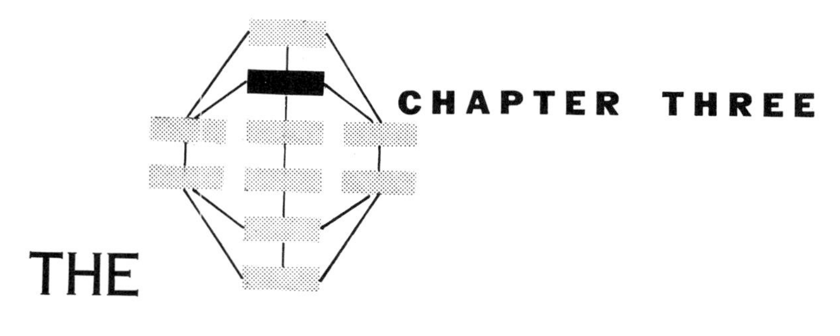

CHAPTER THREE

THE BACTERIAL CELL

In Chapter 1 it was mentioned that the bacteria and the blue-green algae are distinguished from the other protists (and from plants and animals) by the internal organization of the cell. In this chapter the structure of the cell of a typical eubacterium will be considered in detail. Unfortunately, little is known about the fine structure of the blue-green algae or about some of the other kinds of bacteria mentioned in Chapter 1.

As will be pointed out several times in this book, almost the only function of a bacterial cell is to make another cell. All a bacterium does is grow. It does not make cells of different sorts, nor does it do much mechanical work, make seeds and flowers, or walk and talk. In short, it simply makes itself. When we have seen the structure of a bacterial cell, then we will have also seen, nearly in its entirety, the problem such a cell has in reproducing itself. Bacteria have spent several millions of years solving this problem. The bulk of this text is devoted to an exposition of what we know about the solution.

GROSS ANATOMY OF THE BACTERIAL CELL

The Shapes of Bacteria

Real knowledge of bacterial cytology had to await the development of the electron microscope. The conventional light microscope provides not much more than an outline of the bacterial cell and gives only a vague notion of its internal structure. This picture is sufficient, however, to establish the major morphological groups of the eubacteria. Before the fine structure of the bacterial cell is discussed, it is well to learn the variety of forms it may take and something of its gross anatomy.

There are three shapes of cells among the true bacteria: rods, spheres,

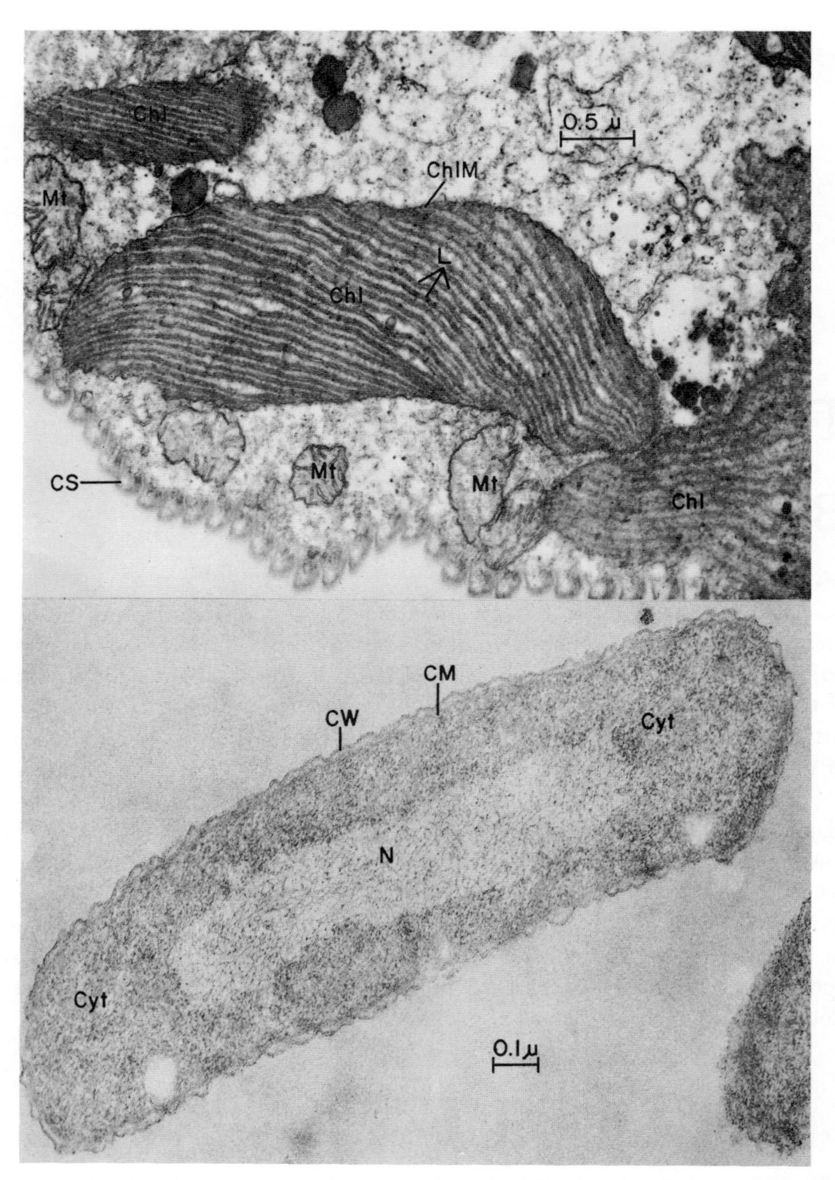

Fig. 3-1. A: An electronmicrograph of part of a cell of *Euglena bacillaris,* a green alga. A portion of the surface of the cell (CS) is at the lower left-hand corner. A large chloroplast (Chl) is in the center of the figure. A membrane (ChlM) surrounds this organelle. The chloroplast is composed of groups of thin, membrane-limited envelopes embedded in a homogeneous matrix material. In the micrograph these envelopes appear as bundles of parallel lines or lamellae (L). Several mitochondria (Mt) are shown. B: A micrograph of *Escherichia coli,* a typical bacterium. The cytoplasm (Cyt) is almost completely structureless. The very small black dots represent ribosomes, the sites of protein synthesis. The cell is surrounded by a rigid cell wall (CW) and a cell membrane (CM). The nucleus (N) is not enclosed by a membrane. (A: courtesy Dr. Sarah P. Gibbs; B: from *Microbial Genetics,* courtesy Dr. E. Kellenberger and Society for General Microbiology.)

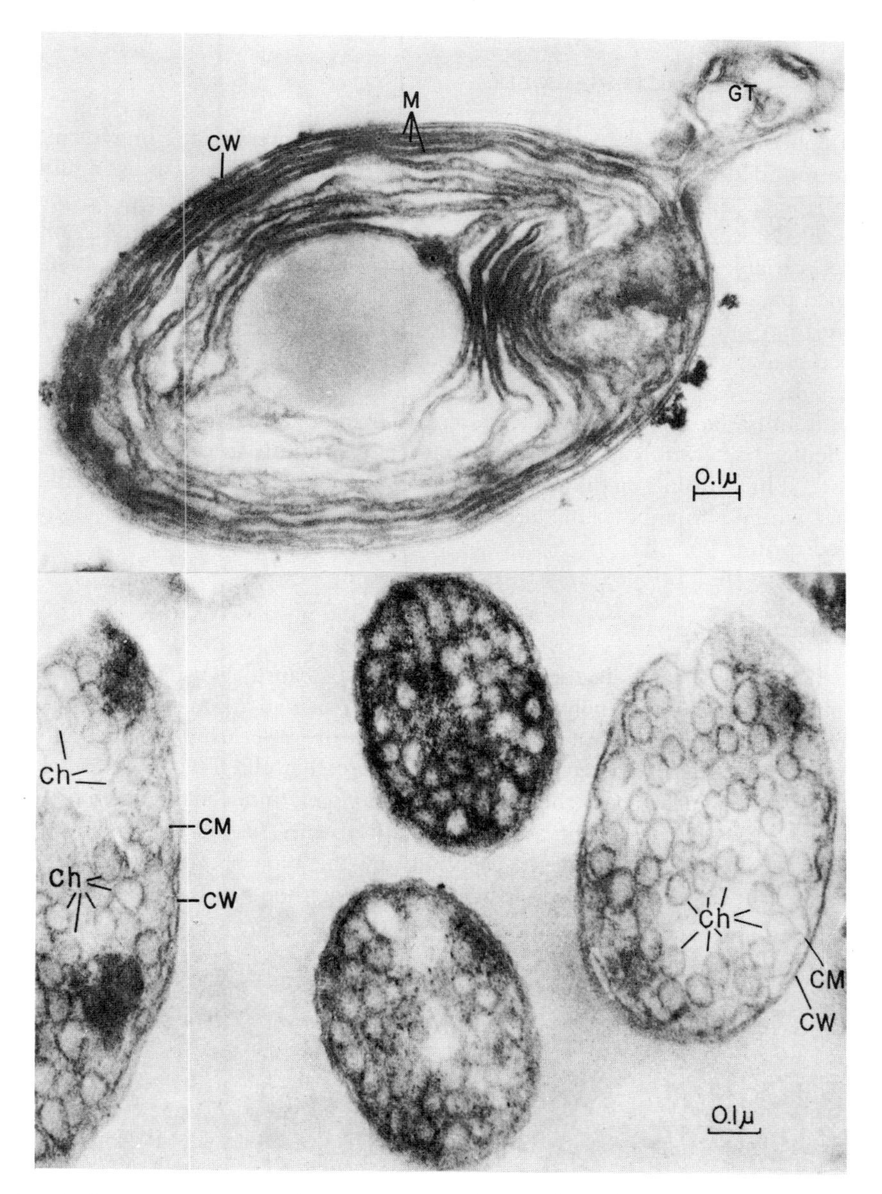

Fig. 3-2. A: An electronmicrograph of the photosynthetic budding bacterium *Rhodomicrobium* (pp. 15 and 51). A small bud or growth tube (GT) is seen at the upper right. The pairs of lines represent sections through packets of membrane-limited envelopes (M); these envelopes are not organized into chloroplasts and the packets are not enclosed in a membrane as in *Euglena*. B: a micrograph of another photosynthetic bacterium, *Rhodospirillum rubrum*. The numerous round structures are chromatophores (Ch). The chloroplast membranes in *Euglena* and the envelopes in *Rhodomicrobium* and the chromatophores in *Rhodospirillum* are the sites of photosynthesis in these organisms; however different in structure, they are similar in function. Notice that the magnification in Fig. 3-1A is only about one third that in the other figures. In other words, an entire cell of *Escherichia coli* is no larger than a chloroplast of *Euglena*. (A, B: courtesy *Journal Biophysical and Biochemical Cytology*, and Dr. H. C. Douglas).

and helices. The rod shape is probably the most common form. The rods may have rounded or blunt ends; some are so short as to be almost indistinguishable from a sphere and others are long and narrow. The spherical bacteria are called cocci. Cocci often occur in chains or in clusters; these clusters may be of variable size and irregular in shape or, in some cases, very regularly shaped packets. The helical forms are called vibrios or spirilla. A spirillum has more than one turn of a helix; a vibrio has less than a full turn and looks superficially like a blunt rod. Some examples of these shapes are shown in Fig. 3-3.

It must be realized that, to a certain extent, the shapes and size of a particular bacterium can vary within rather wide limits in different environments. Thus in one medium a spirillum may have only one or two shallow turns, whereas in another medium this same bacterium can resemble a tightly coiled spring.

The Bacterial Capsule

The cells of many bacteria are surrounded by capsules, usually composed of polysaccharides or of polypeptides of only one or two different amino acids. The capsule is readily demonstrated microscopically, appearing as an envelope around the cell, often much larger than the cell itself. The presence of a capsule is also reflected in the moist, glistening, gelatinous appearance of the colony. The capsule is not a very constant feature, as its size is strongly influenced by the environment, especially by the type of medium. Bacteria can lose the ability to form capsules, usually without suffering any disad-

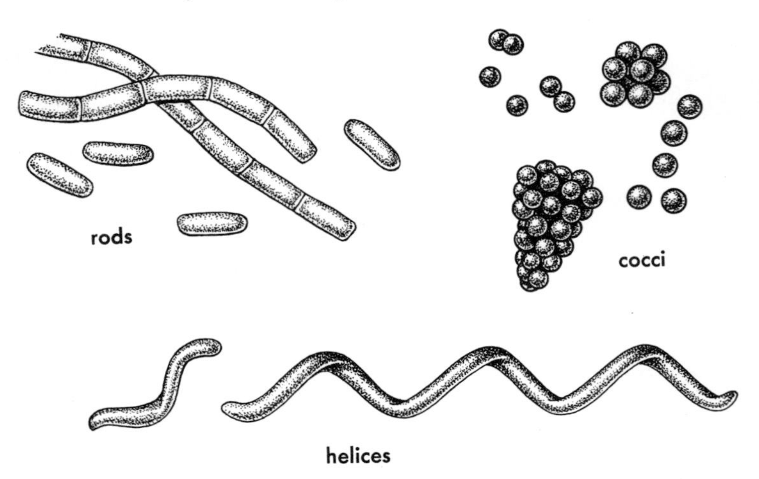

Fig. 3-3. The various shapes the bacterial cell may have are shown in outline. These are cylindrical (rods), spherical (cocci), and helical (spirilla and vibrios).

vantage. The presence of a capsule is of considerable importance in some disease-producing bacteria. For example, the organism that causes pneumonia (Pneumococcus) always possesses a capsule when isolated from a diseased animal. In the laboratory this bacterium can lose the ability to form a capsule; such noncapsulated pneumococci are very easily destroyed by the phagocytes of an infected animal and can no longer cause the disease.

The Gram Reaction

One of the most important cytological features of bacteria is their reaction to a simple staining procedure called, after its discoverer, the Gram stain. The procedure involves staining the cells with the dye crystal violet; all bacteria will be stained blue. The bacteria are next treated with an iodine solution and then decolorized with alcohol; gram-positive bacteria retain the crystal violet, gram-negative bacteria are decolorized. The chemical basis of this differential reaction is not well understood and the procedure appears to be an entirely arbitrary one. However, the division of bacteria into gram-positive and gram-negative forms correlates surprisingly well with many other characteristics. For example, all spore-forming bacteria are gram-positive; all polarly flagellated forms are gram negative. The gram-negative bacteria are as a rule more resistant to many antibiotics than gram-positive ones. Perhaps most importantly, there is a fundamental difference in the structure of the cell wall of the gram-negative and gram-positive bacteria (see p. 28).

Bacterial Flagella

Many bacteria can swim by means of small whip-like appendages called *flagella*. In true bacteria this is the only means of locomotion. The flagellum itself is much too small to be seen under the light microscope. However, it is possible to reveal the shape and position of the flagella by suitable staining procedures and it is easily seen in the electron microscope. The flagella of motile bacteria are distributed over the surface of the cell in characteristic fashions: they may be restricted to one or both ends of the cell (polar flagella), or they may be found all over the cell surface (peritrichous flagella). Peritrichous flagellation is widely distributed among the bacteria, whereas the polarly flagellated bacteria form a rather homogeneous assemblage of rods and spiral-shaped cells. The structure of the flagellum is taken up further on page 24.

The Endospore

The endospore is one of the unique features of the bacteria. Many other microorganisms produce resistant cell types which can remain viable under conditions which would kill the vegetative cells. None of these is the equal of the bacterial spore in withstanding extreme conditions, both chemical and physical. Some endospores can germinate even after several hours' immersion in boiling water. Précisely how the spore manages to survive this kind of treatment is largely unknown and remains one of the most interesting problems in bacterial physiology.

An endospore has a very characteristic appearance when examined microscopically. Because of its low water content it is very dense and highly refractile. Since it does not stain easily, somewhat elaborate means must be employed to stain it. Only one spore is formed in a cell; it may lie in the center or toward one end of the sporangium (the name given to the cell in which a spore is formed). The spore sometimes has a greater diameter than the sporangium, which is spindle- or racquet-shaped in this case.

With only one or two exceptions, spores are restricted to two genera of rod-shaped bacteria: *Bacillus* and *Clostridium.* These genera are distinguished on physiological grounds: the clostridia are strict anaerobes, whereas the bacilli are aerobes (see Chapter 4).

The structure of the endospore varies from organism to organism. Fig. 3-4 is an idealized picture that demonstrates some of the common features of endospores.

Germination of a spore occurs when it is placed in a suitable medium. The spore takes up water and swells, losing its refractility and becoming more easily stained. The spore coats (Fig. 3-4) rupture and the new vegetative cell grows out; the spore wall (Fig. 3-4) forms the wall of the vegetative cell. The remnants of the spore coats may remain attached to the cell but eventually drop off; they form no part of the vegetative cell.

Another kind of resting stage is produced by some bacteria. This stage, called a *cyst,* is much less resistant than the endospore and has a less distinctive morphology. In contrast to the endospore, a cyst is formed from the entire contents of the cell.

Table 3-1 lists some of the more important morphological groups of bacteria, together with a few examples of each.

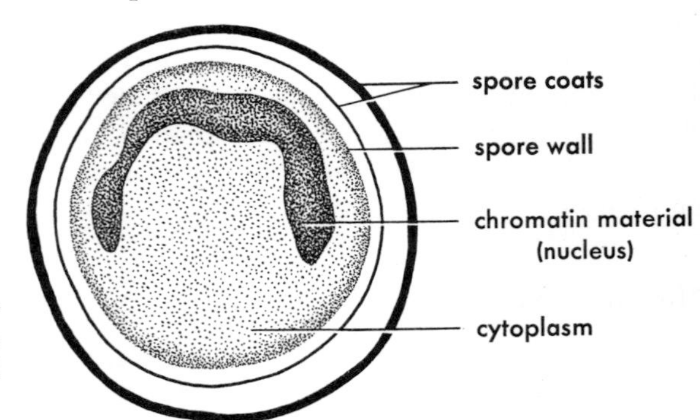

Fig. 3-4. A very diagrammatic representation of a cross section of a typical bacterial endospore.

TABLE 3-1

The Major Morphological Groups of the True Bacteria

Shape	Gram stain	Flagella	Spores	Representative genera
spheres	—	—	—	*Veillonella* *Neisseria*
spheres	+	—	—	*Sarcina, Streptococcus*
spheres	+	single	+	*Sporosarcina*
rods	+	peritrichous	+	*Bacillus* *Clostridium*
rods	+	—	—	*Lactobacillus*
rods	—	peritrichous	—	*Azotobacter* *Escherichia*
rods	—	polar	—	*Pseudomonas*
helices	—	polar	—	*Spirillum*

THE MOLECULAR ARCHITECTURE OF THE BACTERIAL CELL

The Chemistry of the Cell

One way to look at the cytology of a bacterium is by an analysis of its chemical composition. As it does in all living things, water makes up the bulk of a bacterial cell: 70 to 80 percent of the weight of a cell is water. A typical bacterium weighs 5 times 10^{-6} μg[1] and has a dry mass of about 1 times 10^{-6} μg. What is the chemical composition of this one trillionth of a gram? Its composition is unremarkable: the common elements carbon, hydrogen, oxygen, and nitrogen, together with lesser amounts of phosphorus and sulfur, comprise at least 99 percent of the total dry mass. There are smaller amounts of iron, potassium, magnesium, and some other elements. Typical values are shown in Table 3-2.

An elemental analysis of this sort is not, of course, very useful except in the negative way of telling us that significant information can come only from a molecular analysis. Table 3-3, which shows some values of a more useful kind of analysis, gives a picture of the molecular building blocks that go to make up the cell.

You will see that there is nothing very remarkable about the bacterial

[1] μg or microgram. A microgram is one one-thousanth of a milligram, which is one one-thousandth of a gram.

TABLE 3-2

Composition of a Typical Bacterium

	percent dry weight
carbon	50
oxygen	20
nitrogen	15
hydrogen	8
sulfur	3
phosphorus	1

cell as far as its macromolecular constituents are concerned. The most striking feature is probably the rather high concentration of ribonucleic acid, which will be discussed in Chapter 6. In the rest of this chapter we shall attempt to see how these various components are distributed structurally within the cell.

The proteins, nucleic acids, lipids, and polysaccharides of the bacteria have the same molecular architecture as they do in animals or plants. The important structural features of these molecules have already been treated in *Cell Structure and Function,* in this series, and need not be described here. However, the reader is advised to have these features firmly in mind.

It will be recalled that these macromolecules are composed of many smaller molecules linked together. For example, proteins are made from amino acids. The small molecules can be considered building blocks. As will be shown later, the bacterial cell synthesizes its macromolecular constituents in two steps: first it makes (or acquires) the small molecules or building blocks, and then it puts them together to form the macromolecules. The building blocks are the same in all bacteria and indeed the identical building blocks are used throughout the living world. It is in the structure of the macromolecules that biological specificity resides and it is in the synthesis of these molecules that the cell determines itself.

Between the picture given by the light microscope and that revealed by chemical analysis lies the most interesting area of bacterial cytology. If we think of a cell as a city, then the macromolecules would be the buildings and the small molecules would be the bricks and concrete and nails. In this image the interesting features would be the neighborhoods. The details of these features of bacterial anatomy can be revealed only by the combined use of the electron microscope and chemical analysis.

Structure of the Bacterial Flagellum

A typical bacterial flagellum is about 120A (Ångstrom units) thick and may be 4 or 5 μ long. The flagellum is attached to a small basal granule inside the cell and protrudes through the cell wall. The flagellum is cytoplasmic in origin and is not an appendage of the cell wall. It is helical in shape, with a constant pitch and radius for a given type of bacterium. In a dried preparation under the microscope, the helix is flattened out and the flagellum appears as a wavy line.

It is easy to obtain a pure preparation of flagella. Analysis of purified flagella shows that they contain only protein, and that the protein is similar in amino acid composition to the myosin from muscles of higher organisms. Recalling that the thickness of one helix of protein is about 10A, it is clear that a flagellum (120A thick) is composed of only a very few protein molecules. A bacterial flagellum does not show very much structural detail, even under the electron microscope. The flagellum seems to be composed of two (or three) strands twisted around one another. The cilia and flagella of all other organisms are composed of two central strands of protein surrounded by nine peripheral strands (each of these strands is not much thicker than the entire flagellum of a bacterium). That this pattern is not evident in bacterial flagella is one of the features distinguishing the bacteria from all other organisms.

TABLE 3-3

Molecular Composition of a Bacterium

Substance	Mass times $10^7 \mu g$	Molecular weight[a]	Molecules per cell	Molecular abundance
total mass	50	—	—	—
water	40	18	1.3×10^{10}	5×10^{6}
dry mass	10	—	—	—
protein				
ribosomal protein	1	1×10^{5}	$.6 \times 10^{6}$	250
other protein	4	1×10^{5}	2.4×10^{6}	950
total protein	5		3.2×10^{6}	1000
ribonucleic acid				
ribosomal	1.2	1×10^{6}	8×10^{4}	32
soluble	.3	1×10^{6}	2×10^{4}	8
total	1.5		10×10^{4}	40
deoxyribonucleic acid	.2	5×10^{6}	2.5×10^{3}	**1**
carbohydrate (polysaccharide)	1	5×10^{2}	3×10^{8}	1.2×10^{5}
lipid	1	1×10^{3}	1.5×10^{8}	$.6 \times 10^{5}$

[a] These molecular weights are rough averages.

The flagella drive the bacterium forward in much the same way as a screw propels a ship. However, unlike the propeller of a ship, the flagella do not themselves rotate; rather, screw-shaped waves move down them in the same way as a wave will pass down a rope tied at one end and rotated at the other. In peritrichously flagellated organisms the flagella form a bundle or tuft which extends back from the posterior end of the cell. This tuft of flagella is the actual propulsive organ and is thick enough to be seen in the light microscope under certain conditions. Many polarly flagellated organisms also have a tuft of flagella at the pole, whereas other polarly flagellated cells seemingly have only a single flagellum.

The Cell Wall and Cell Membrane

Before considering the molecular structure of the two limiting envelopes of the bacterial cell—the wall and the membrane—it is necessary to review briefly the phenomenon of osmosis.

Consider a solution of sucrose in water. The molecules of both sucrose and water are in violent and random motion throughout the solution. If the solution is placed in a bag through which sucrose cannot pass but through which water can pass (the bag is said to be impermeable to sucrose but permeable to water) and the bag is placed in pure water, what will happen? The water molecules will pass through the bag in both directions; but on the average more will go in than out. This is because, for a given total *number* of molecules, there are fewer water molecules inside the bag than outside since some of the molecules inside are sucrose. In a closed bag which is full, the solution will be under a pressure; this is called the osmotic pressure and is a measure of the tendency of the water to pass into the bag. If the bag is elastic, it will swell and eventually burst if the osmotic pressure becomes greater than the mechanical strength of the bag. On the other hand if a bag full of pure water is placed in a solution of sucrose, it will shrink as the water flows out of it.

The same kind of behavior is shown by bacteria under similar circumstances. Thus, if a bacterial cell is placed in a very concentrated solution of sucrose, the cell contents (the protoplast) can be seen to shrink, but the outer bounding envelope does *not* shrink, it is rigid. This very simple experiment, performed more than 50 years ago but ignored for a long time, demonstrates two important properties of the surface of the bacterial cell. First, there is a semipermeable membrane around the cell, since the cell contents respond to a change in the osmotic pressure of the medium. Second, on the outside of the membrane there is a *rigid cell wall*. Some idea of the rigidity of this wall may be gained by considering the osmotic pressure exerted against the wall when a bacterium is placed in water (treatment which usually causes no great harm or visible change). It is known that the osmotic pressure is then about 20 atmospheres or 300 pounds per square inch. The membrane responsible for the osmotic behavior is called the *plasma membrane* and the rigid structure which resists the osmotic pressure, the *cell wall*. These are functional definitions of the membrane and cell wall; it is only recently that microbiologists have been able to study separately the structures responsible for these two functions. It is possible to remove the cell wall of some bacteria leaving the protoplast intact. The protoplast is bounded by the plasma membrane. Since it is no longer constrained by a rigid wall, the protoplast is spherical. For the same reason, it is very sensitive to changes in the osmotic pressure of the medium. Protoplasts swell and burst if placed in water and shrink if placed in a solution whose osmotic pressure is greater than their own. Protoplasts can carry out all the metabolic activities of the whole cell, thus demonstrating that the only function of the cell wall is a mechanical one.

The Structure of the Bacterial Cell Wall

What is the chemical nature and structure of this rigid cell wall? Many other organisms have rigid walls. In higher plants and in some algae the wall is composed of cellulose while in the fungi it is largely chitin. Both cellulose and chitin are polysaccharides. The cell walls of bacteria also contain polysaccharides as major components, but not chitin or cellulose.

The cell wall of a bacterium is easily seen in the electron microscope (see Fig. 3-2b). It usually appears as an envelope 50–100A thick. To learn more about the structure of the cell wall we must investigate its chemical composition.

Only in recent years have microbiologists been able to determine the chemical composition of the cell wall. The first step was the development of methods for obtaining large amounts of purified cell wall material which could be subjected to chemical analysis.

Polysaccharides constitute a large fraction of the cell wall. The polysaccharides are composed of a number of different sugars, such as glucose, galactose, and mannose. The sugars found depend on the bacterium. In many cases amino sugars occur rather than the corresponding simple sugars. The difference between these can be seen by comparing the structures of galactose and galactosamine.

$$\begin{array}{c} \mathrm{CHO} \\ | \\ \mathrm{HCOH} \\ | \\ \mathrm{HOCH} \\ | \\ \mathrm{HOCH} \\ | \\ \mathrm{HCOH} \\ | \\ \mathrm{CH_2OH} \\ \text{galactose} \end{array} \qquad \begin{array}{c} \mathrm{CHO} \\ | \\ \mathrm{HCNH_2} \\ | \\ \mathrm{HOCH} \\ | \\ \mathrm{HOCH} \\ | \\ \mathrm{HCOH} \\ | \\ \mathrm{CH_2OH} \\ \text{galactosamine} \end{array}$$

An almost universally present component of the polysaccharides of bacterial cell walls is the compound muramic acid, which is related to galactosamine.

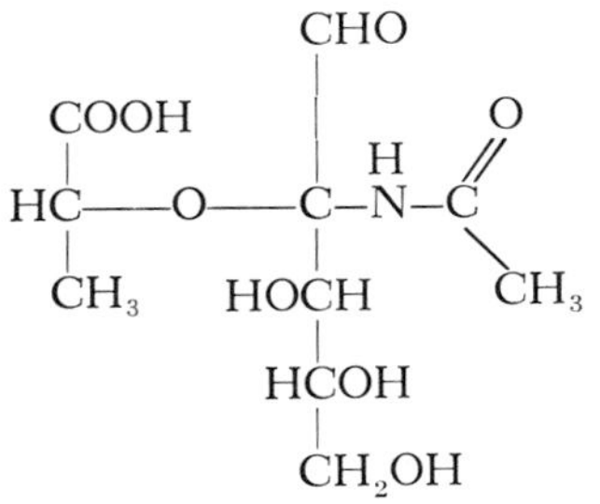

```
                    CHO
                     |
   COOH              |          O
    |                |   H    //
   HC------O--------C--N--C
    |                |        \
   CH3           HOCH          CH3
                     |
                   HCOH
                     |
                   CH2OH
```

muramic acid

Muramic acid has so far been found only in the bacteria and actinomycetes. This is one reason for believing these two groups to be closely related.

The wall of gram-positive bacteria contains only traces of lipids, while the wall of gram-negative bacteria may contain up to 20 percent lipid. This is one of the differences between the cell walls of the two groups of bacteria. Another, more striking difference is seen in the amino acid composition of the walls. The gram-negative wall has a wide range of amino acids; indeed all the amino acids which occur in proteins are found in the cell walls of gram-negative bacteria. The gram-positive wall, on the other hand, is characterized by the striking simplicity of its amino acid composition. Glutamic acid and alanine usually occur in large amounts, accompanied by smaller amounts of one or two other amino acids which are different in different bacteria. The amino acid diaminopimelic acid occurs in the walls of both gram-positive and gram-negative bacteria. Like muramic acid, diaminopimelic acid is found only in the lower protists (bacteria and blue-green algae), and is confined almost entirely to the cell wall. The limited distribution of these two compounds is yet another indication of the separateness of these organisms.

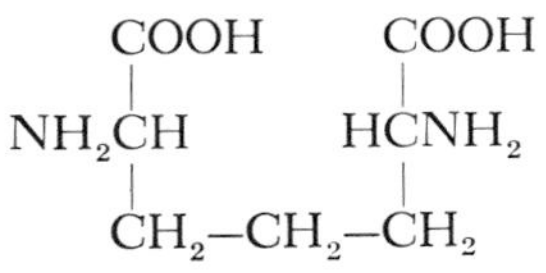

```
      COOH          COOH
       |             |
   NH2CH           HCNH2
       |             |
      CH2--CH2--CH2
```

diaminopimelic acid

Clearly, the structure of the wall in the gram-positive bacteria is simpler than it is in the gram-negative bacteria. Although the structure of the cell wall in the gram-positive bacteria certainly shows some variation, it appears to be based on a simple plan exemplified by the structure shown in Fig. 3-5.

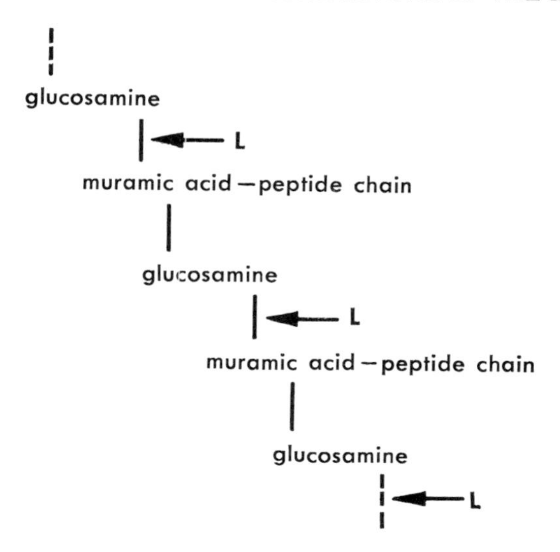

Fig. 3-5. The basic component of the bacterial cell wall. The polysaccharide forms a large fraction of the cell wall of gram-positive bacteria and is also found in gram-negative bacteria. The enzyme lysozyme attacks the bonds between galactosamine and muramic acid, marked L in the figure.

The backbone of this structure is a polysaccharide composed, in this case, of glucosamine and muramic acid. The carboxyl group (COOH) of muramic acid is connected to a short chain of amino acids by a peptide bond.

As mentioned previously, it is possible to remove the cell wall of some gram-positive bacteria without damaging the protoplast within it. One way this can be done is by use of the enzyme *lysozyme,* which hydrolizes certain kinds of polysaccharides. Lysozyme attacks the bond between glucosamine and muramic acid, as shown in Fig. 3-5. The walls of some gram-positive bacteria are completely digested by lysozyme; hence, they must be composed entirely of a polysaccharide-peptide complex very similar to the one shown in Fig. 3-5. However, the walls of other gram-positive bacteria are digested only partially. In these cases the wall must have a somewhat different structure.

It has already been shown that the cell wall of gram-negative bacteria is chemically very much more complex than that of gram-positive bacteria. It is also architecturally more complex. The rigidity of the gram-negative wall is due to a component very similar in structure to the entire gram-positive wall. Over this rigid layer are two or three layers of lipopolysaccharides and of proteins. A close inspection of Fig. 3-2b will show that the wall of *Escherichia coli* is composed of at least two layers.

The plasma membrane

Unfortunately, knowledge of the plasma membrane is very limited. A chemical analysis is not very helpful because of the membrane's complexity. The membrane contains large amounts of phospholipids together with proteins and some polysaccharides. From electron micrographs it appears to be not more than 20–40A thick, thus indicating that its protein molecules are arranged in a single layer.

The most well-established fact about the bacterial plasma membrane is that the respiratory enzymes of the cell are associated with it. In your study of the structure of cells of other organisms you learned that the respiratory enzymes are confined to organelles called mitochondria; several mitochondria can be seen in the alga shown in Fig. 3-1. Examination of the micrograph of the bacterium in Fig. 3-2 shows that, as in the case of all bacteria, there are no structures resembling mitochondria in this cell. The greater simplicity in the organization of the bacterial cell is a result of its small size. In bacteria, as in other cells, many enzymes including the respiratory enzymes are bound to membranes. The amounts of such enzymes a cell can have will be determined by the area of the membranes. In a large cell the area of the cell membrane is not sufficient to accommodate as many enzyme molecules as the size of the cell requires. Structures such as mitochondria clearly increase the surface area enormously. But, as was mentioned in Chapter 1, the smaller a cell is, the larger the ratio of surface area to volume. Bacteria, which are about the same size as mitochondria, do not need "micromitochondria" within them. In some bacteria at least, the plasma membrane extends itself into the cytoplasm in short loops or folds. Perhaps we see here the earliest stages in the development of internal membranes.

The Cytoplasm and Nucleus

To the student used to the wonderful complexity of plant and animal cells, the absence of structure in a bacterial cell is perhaps the most noticeable feature of its cytology. The cell shown in Fig. 3-2b does not seem to offer much scope to the cytologist. This cell, however, can make an exact copy of itself in half an hour or so. The fact that the bacterial cell has a simple structure means then that the complexity of the cells of higher organisms is not essential to growth. But what *is* visible in this cell? The central light area, containing the deoxyribonucleic acid of the cell, can be equated functionally with the nucleus of other cell types. Note the absence of a nuclear membrane. As far as it is known, bacteria do not have chromosomes; that is, the nuclear region does not contain smaller structural units of constant morphology. Since they lack chromosomes, bacteria do not perform the intricate choreography of mitosis at cell division. Very little is known concerning the division of the bacterial nuclear region; apparently it simply pinches itself into two halves.

The space between the nuclear region and the plasma membrane is filled with a uniform, granular cytoplasm. On a closer examination of the Fig. 3-2b, innumerable very small particles can be seen in the cytoplasm. These particles are ribosomes; they are approximately 100A in diameter. The ribosomes of bacteria are apparently free in the cytoplasm and are not associated with any membrane component as they so often are in other cells. The ribosomes consist of about equal amounts of protein and ribonucleic acid. The RNA of the ribosomes represents 80 to 85 percent of all the RNA of the cell. The remaining 15 to 20 percent is dissolved in the cytoplasm. This is the soluble RNA in Table 3-3. The protein of the ribosomes is part of the apparatus for making other proteins. A great deal of evidence suggests that ribosomes are the seat of protein synthesis, which will be covered more fully in Chapter 6.

Deposits of various substances, most often lipids, very frequently occur in bacteria. These inclusions can easily be seen under the light microscope. Since the number and size of these inclusions depend greatly upon the environment of the cell, they do not represent constant or essential features of the cellular organization.

THE MECHANICS OF CELL DIVISION

Bacteria reproduce by dividing in half; the division is perpendicular to the longer axis of the cell in rods and spirilla. This process is called simple binary fission, but the word "simple" probably reflects our small knowledge of the process rather than any real simplicity of the process itself.

The division of the nucleus has already been described; the division of the cytoplasm occurs in several stages. First, a transverse plasma membrane is laid down. This membrane is then split in half by the centripetal growth of the cell wall. Very often this transverse cell wall will remain incomplete, with the two daughter cells connected at the transverse membrane. The two daughter cells may remain together and continued division leads to a chain or a cluster of cells. In the spherical bacteria (cocci) the planes of successive division may be parallel producing chains), or perpendicular to one another, producing sheets or packets of cells. Some cocci characteristically have a very regular succession of the division planes and so form cuboidal packets of eight or more cells.

THE CYTOLOGY OF THE PHOTOSYNTHETIC LOWER PROTISTS

In all photosynthetic organisms except the blue-green algae and photosynthetic bacteria, the essential steps of photosynthesis are carried out in well-defined organelles called chloroplasts. The chloroplast is a complex structure that is composed essentially of layers of membranes and is delimited by a membrane. A typical chloroplast is shown in Fig. 3-1. The details of the structure of this organelle can be found in *The Living Plant,* in this series. Two photosynthetic bacteria are shown in Fig. 3-2a. The first point to be noticed is that these bacteria are themselves smaller than the chloroplast shown in Fig. 3-1. Second, the organization of the photosynthetic apparatus is entirely different from that seen in the alga. The photosynthetic pigments of these bacteria are contained in chromatophores. In one case, the chromatophores are simple round bodies; in the other, the chromatophores are more complex and consist of four or five lamellae packed together. In neither case is there a membrane surrounding the whole structure. It must be realized that the bacterial chromatophore is not an incomplete chloroplast. Isolated chromatophores can carry out many of the reactions of photosynthesis. The absence of an extensive system of internal membranes in the photosynthetic lower Protista is analogous to the absence of mitochondria. The photosynthetic apparatus of the blue-green algae apparently consists of isolated membranes that are more or less irregularly distributed throughout the cell. Here again, there is no chloroplast membrane separating the lamellae from the rest of the cell.

CHAPTER FOUR

ENERGY METABOLISM

The next two chapters are devoted to the metabolism of bacteria. This chapter is concerned with the energy metabolism and Chapter 5 with the biosynthesis of the building blocks needed for growth. Before these two kinds of metabolism are examined in detail, it is well to consider them broadly in the light of the general economy of a bacterial cell.

THE NEED FOR ENERGY

Like any organism, a bacterium exists only to make a copy of itself; in other words, a bacterial cell is a "self-replicating unit." If we analyze the meaning of this term, we see that there are really two ideas involved. In the first place, there is the notion of an increase in the number of a specific kind of unit. This, of course, is what replication means in any context. It is when we come to the idea of *self*-replication that we approach peculiarly biological problems. The copies of this book have been replicated, but they did not self-replicate. Each of these pages has a specific pattern that is repeated in every copy of a page, but this specificity was imposed from without. The pattern was in the printer's type, which replicated pages but did not replicate itself. A bacterial cell (or any organism) is a copy of itself, and the specificity is contained within the unit which is replicated.

A cell is self-replicating, but clearly it must obtain from the environment all the material it needs to replicate itself. To change the material of the environment into a copy of itself, a cell must also have a supply of energy.

Why does a cell require energy simply to grow? When it is just growing, a cell is not moving about, or pumping blood, or lifting weights. In fact it is not doing any of the things one might intuitively associate with work. A growing cell needs energy because it is making order from disorder; it is making the complex and highly ordered structure of a cell out of the disorder of the environment. It is a basic principle of physics that the creation of order does not proceed *spontaneously*. Rather, the reverse is true; that is, disorder comes from order spontaneously. Therefore, work must be done, or, in other words, energy must be expended to form an ordered system. The kind of order a cell makes when it replicates itself is chemical. The macromolecular components of the cell (proteins, nucleic acids, and polysaccharides) are very precisely ordered molecules, and they are themselves parts of an ordered structure. Furthermore, the building blocks which comprise these large molecules must be made by the cell or obtained ready-made from the environment. These processes require the expenditure of chemical energy.

Within a growing cell then we can distinguish, on the basis of their functions, two metabolisms. One is biosynthetic, and includes all the reactions by which material from the environment in transformed into building blocks and specific macromolecular components of the cell. The second metabolism supplies the energy consumed in the first. It must not be supposed that these two metabolisms are really separate. On the contrary, many reactions and intermediates are common to both. Thus, the functions are separable but the mechanisms are closely interwoven. It is for this reason that we have avoided using "catabolism" (or breaking down) to mean "energy metabolism" and "anabolism" (or building up) to mean biosynthesis.

The link between the reactions which consume energy and those which supply it is provided by adenosine triphosphate (ATP). The functions of this all-important compound are described in *Cell Structure and Function* in this series. The formation of ATP is the mechanism by which the energy available from any metabolic reaction can be captured, and in turn used to drive any reaction requiring energy. Briefly, the way ATP works is as follows. The formation of ATP from adenosine diphosphate and phosphoric acid is an endergonic reaction; that is, it consumes energy. This reaction can proceed only if it is coupled to an energy-yielding or exergonic reaction. In this way the energy from various exergonic reactions, which would otherwise simply produce heat, is captured in a single compound. The energy stored in ATP can be used to drive the multitude of endergonic biosynthetic reactions necessary for growth.

In the economy of the cell, ATP can be likened to money in that it can be made and spent in a variety of ways. Without money, an economy is based on barter, and without ATP each different energy-consuming reaction would require a special link to an energy-yielding reaction. It may be noted in passing that ATP, unlike money, cannot be created without an expenditure of energy. In the course of their evolution, bacteria have developed an enormous variety of energy metabolisms. The purpose of *all* of these is to provide the cell with energy in the form of ATP.

OXIDATIONS AS SOURCES OF ENERGY

Ultimately only two sources of energy are available to living things: sunlight and chemical oxidation. Organisms which use sunlight are called *phototrophs;* those which use chemical energy are called *chemotrophs.* The phototrophic bacteria will be treated later on in this chapter; for the moment we shall confine our attention to the chemotrophs.

First we must digress briefly to consider the salient features of oxidation–reduction reactions. Oxidation is the removal of electrons from an atom or molecule; reduction is the addition of electrons. The oxidation of iron is an example of the simplest kind of oxidation, that in which a single electron is removed:

$$Fe^{2+} \rightarrow Fe^{3+} + e^-.$$

The oxidation of organic compounds, especially those of biological importance, most often involves the removal not of free electrons but of hydrogen atoms, usually in pairs. This is called dehydrogenation. The oxidation of succinic acid to fumaric acid is an example:

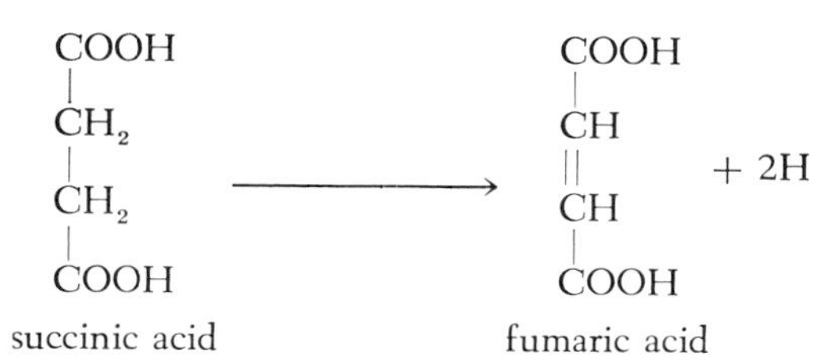

succinic acid fumaric acid

Neither electrons nor hydrogen atoms can accumulate as such; therefore, every oxidation is accompanied by a reduction. Thus the oxidation of succinic acid, just illustrated, cannot occur unless another compound is present which can accept the hydrogen atoms. The hydrogen acceptor is thereby reduced. Any oxidation can be represented by the general reaction

$$AH_2 \rightarrow A + 2H$$

and a reduction by

$$B + 2H \rightarrow BH_2$$

The sum of the two reactions represents the oxidation of AH_2 by B:

$$AH_2 + B \rightarrow BH_2 + A$$

In this reaction AH_2 is the reductant (or hydrogen donor) and B is the oxidant (or hydrogen acceptor).

All the energy-yielding reactions of chemotrophic bacteria are oxidation–reductions and involve the transfer of hydrogen atoms (or electrons) from one compound to another. The maximum amount of energy the cell can obtain from the transfer of hydrogen atoms from one compound to another cannot exceed the amount of energy released when the same transfer occurs outside the living cell.

To be of use to an organism as an energy source, a particular reaction must involve an oxidant and a reductant that are present in sufficiently large amounts in the environment. There are, in fact, a large number of such reactions, and it is remarkable that the bacteria as a group have evolved mechanisms for making use of nearly all of them as energy sources.

The energy-yielding oxidations of bacteria may be classified according to the organic or inorganic nature of both the oxidant and the reductant, as shown in Table 4-1 (facing).

The terms *lithotroph* and *organotroph* are new to you. Lithotrophs, which are found only among the bacteria, are organisms that obtain their energy from the oxidation of inorganic reductants, such as sulfur and iron, by inorganic oxidants. Organotrophs, in contrast, oxidize organic hydrogen donors, using either organic or inorganic oxidants. All animals, the fungi, and most bacteria are organotrophs.

The names respiration and fermentation are probably already familiar, although with more restricted meanings than are used here. Respiration is defined here as any energy-yielding oxidation in which the oxidant is an inorganic compound; oxygen need not be involved. In a similar way, fermentation is defined as any energy-yielding oxidation in which the oxidant is organic.

A detailed treatment of these several classes of energy metabolism will now be given. It should be borne in mind that as far as the bacteria are concerned, the point of these reactions is simply to obtain the ATP needed for growth.

FERMENTATION

Fermentations of Carbohydrates

Carbohydrates, including polysaccharides, are the most abundant and important substrates of bacterial fermentations. The fermentations of the simple sugar glucose are typical, since the fermentations of other sugars differ only in detail and since polysaccharides are always hydrolyzed to the constituent sugars before being fermented.

At least seven distinct fermentations of glucose occur among the bacteria. Each is associated with a specific set of end products, and each is characteristic of a particular group of bacteria. Several of these fermentations are of economic importance. Here the emphasis will be on their similarities and their role in energy metabolism rather than on the formation of the individual end products.

TABLE 4-1
Energy-Yielding Oxidations

Hydrogen donor	Hydrogen acceptor	
	Inorganic	Organic
inorganic (lithotrophic)	respiration	does not occur
organic (organotrophic)	respiration	fermentation

All of these fermentations can be considered as proceeding in two stages. The first stage, the conversion of glucose to pyruvic acid, involves the splitting of the carbon chain of glucose and the removal of two pairs of hydrogen atoms. This is the oxidative half of the fermentation. In the second stage, the hydrogen atoms from the first stage are used to reduce either pyruvic acid or compounds derived from pyruvic acid.

Three different pathways leading from glucose to pyruvic acid are known in the bacteria. These are outlined in Fig. 4-1. Pathway *A* is the glycolytic pathway found in yeast and muscle. This is known as the Embden-Meyerhof-Parnas (EMP) scheme after its discoverers. Pathway *B*, known as the Entner-Doudoroff (ED) scheme, has been found only in bacteria. Finally, pathway *C*, the hexose monophosphate (HMP) scheme, occurs in a large number of organisms.

The details of the EMP pathway are described in *Cell Structure and Function* in this series. This pathway is composed of a number of simple reactions, each catalyzed by a specific enzyme. The initial reactions lead from glucose to fructose diphosphate. The formation of this compound is characteristic of glycolysis and does not occur in either of the other two pathways. The next step is the splitting of fructose diphosphate to yield two molecules of glyceraldehyde phosphate. This reaction also is unique to glycolysis. The dehydrogenation of glyceraldehyde phosphate that follows is the energy-yielding oxidation in this pathway. The dehydration is catalyzed by glyceraldehyde phosphate dehydrogenase. The hydrogen atoms are not transferred directly to the final acceptor but to diphosphopyridine nucleotide (DPN). Since DPN is present in the cell in very small amounts, the fermentation can continue only if the reduced DPN is reoxidized. This reoxidation is accomplished in the second stage of the fermentation, wherein reduced DPN transfers the hydrogen to the final hydrogen acceptor. DPN plays this role of hydrogen carrier in nearly all fermentations. The EMP path is shown in Fig. 4-2.

The energy released in the oxidation of glyceraldehyde phosphate is

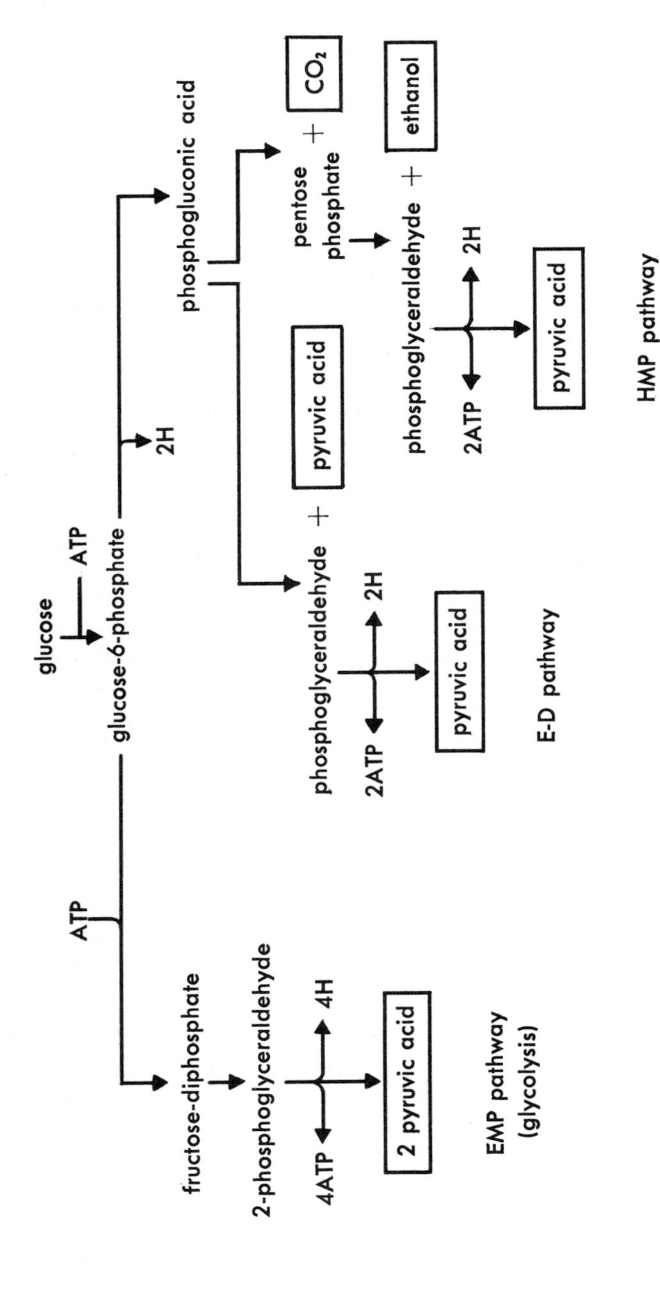

Fig. 4-1. The pathways of glucose metabolism. The figure shows, in barest outline, the three known pathways for the conversion of glucose to pyruvic acid.

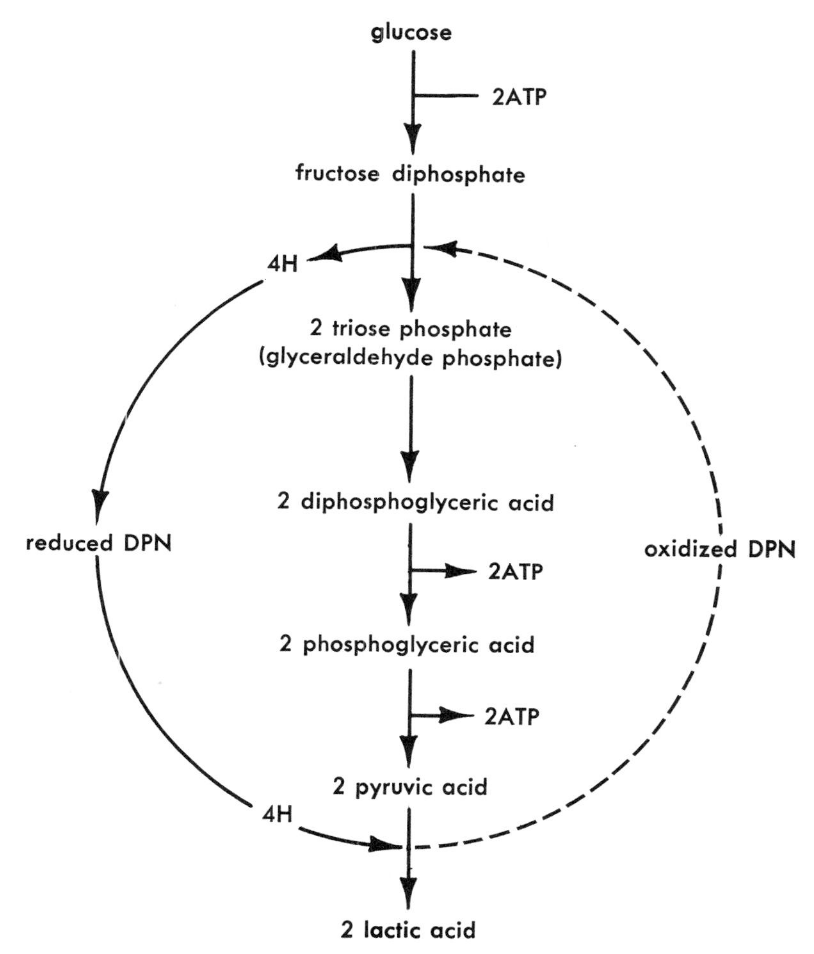

Fig. 4-2. Glycolysis in the lactic acid bacteria. In this figure the scheme for glycolysis has been much simplified in order to emphasize hydrogen transport and the generation of ATP. Further details can be found in *Cell Structure and Function,* in this series.

sufficient for the formation of two molecules of ATP. Since each molecule of glucose yields two of glyceraldehyde phosphate, a total of four molecules of ATP are formed. However, because two of these are required to convert glucose to fructose diphosphate, only two ATP molecules remain for use in growth.

As can be seen from Fig. 4-1, the two other pathways differ in several respects from glycolysis. The most important difference is that in each of them only one molecule of ATP is formed for each molecule of glucose fermented. In all three pathways it is the oxidation of glyceraldehyde phosphate that leads to the formation of ATP. In glycolysis, each molecule of glucose yields two of glyceraldehyde phosphate. In the other pathways, one pair of hydrogen atoms is removed before glucose is split, and therefore only one molecule of glyceraldehyde phosphate is formed. As in glycolysis, the oxidation of glyceraldehyde phosphate is coupled to the formation of two ATP. Inasmuch as only one ATP is required in the formation of phosphogluconic acid, the net yield is one ATP for every glucose fermented.

The HMP pathway differs from the other two in that only one molecule of pyruvic acid is formed, along with CO_2 and ethanol. These last two compounds are among the end products wherever this pathway is involved. In other respects the nature of the end products formed during the fermentation of glucose does not depend upon which of these pathways is operative.

The characteristic end products are derived from the pyruvic acid and the hydrogen atoms produced in the first stage of fermentation. These products are derived from pyruvic acid by sequences of reactions catalyzed by specific enzymes. The end products characteristic of a bacterium depend then upon the particular array of these enzymes it possesses.

Two examples may help to clarify these ideas. Yeast ferments glucose via the glycolytic pathway to ethanol and carbon dioxide. Yeast cells contain pyruvic decarboxylase, which catalyzes the following reaction:

$$\underset{\text{pyruvic acid}}{CH_3COCOOH} \rightarrow \underset{\text{acetaldehyde}}{CH_3C\begin{matrix}\diagup H\\ \diagdown\!\!\!\diagdown O\end{matrix}} + CO_2$$

Ethanol arises from the reduction of acetaldehyde by the reduced DPN formed in the oxidation of glyceraldehyde phosphate. In other words, in this fermentation acetaldehyde is the final hydrogen acceptor.

$$CH_3C\begin{matrix}\diagup H\\ \diagdown\!\!\!\diagdown O\end{matrix} + 2H \rightarrow \underset{\text{ethanol}}{CH_3CH_2OH}$$

The overall fermentation is represented by the following equations:

$$\underset{\text{glucose}}{C_6H_{12}O_6} \xrightarrow{\qquad\qquad} \underset{\text{pyruvic acid}}{2CH_3COOOH} + 4H$$

$$2CH_3COCOOH + 4H \rightarrow 2CH_3CH_2OH + 2CO_2$$

The simplest possible fermentation occurs in certain bacteria in which pyruvic acid itself is the final hydrogen acceptor. The reduction of pyruvic acid produces lactic acid.

$$CH_3COCOOH + 2H \rightarrow CH_3CHOHCOOH$$

If the pyruvic acid is formed via the glycolytic pathway, lactic acid is the sole product of the fermentation; the overall equation is then

$$C_6H_{12}O_6 \rightarrow 2CH_3CHOHCOOH$$

In these two examples the reactions of pyruvic acid serve only to provide the acceptors for the hydrogen atoms arising from the oxidation of glyceraldehyde phosphate; no more ATP is formed. In many bacteria the reactions of the second stage of fermentation involve an additional oxidation and the formation of more ATP. The actual yield of ATP from a complete fermentation may then be greater than that from the first stage alone.

Fermentation of Amino Acids

Many compounds, in addition to carbohydrates, are fermented by various bacteria. An example is the fermentation of amino acids by clostridia. The clostridia are a group of gram-positive, spore-forming rods. A number of clostridia can hydrolyze proteins to amino acids, which they then ferment. In these fermentations the amino acids are often fermented in pairs, one amino acid serving as the oxidant and the other as the reductant. Thus a mixture of alanine and glycine is fermented to acetic acid and carbon dioxide. The alanine is oxidized to pyruvic acid which is itself further oxidized to acetic acid and carbon dioxide.

$$\underset{\text{alanine}}{CH_3\underset{\displaystyle NH_2}{\underset{|}{C}}HCOOH} \rightarrow 2H + NH_3 + \underset{\text{pyruvic acid}}{CH_3COCOOH} \rightarrow \underset{\text{acetic acid}}{CH_3COOH} + 2H + CO_2$$

Glycine serves as the acceptor for the hydrogen atoms from these two oxidations and is reduced to acetic acid.

$$\underset{\text{glycine}}{H_2NCH_2COOH} + 2H \rightarrow CH_3COOH + NH_3$$

The oxidation of one molecule of alanine produces two pairs of hydrogen atoms, whereas the reduction of glycine requires only one pair. In this fermentation, therefore, twice as much glycine is used as alanine.

$$\text{alanine} + 2H_2O \rightarrow \text{acetic acid} + CO_2 + NH_3 + 4H$$

$$2\ \text{glycine} + 4H \rightarrow 2\ \text{acetic acid} + 2NH_3$$

The overall equation for the fermentation is then

$$\text{alanine} + 2\ \text{glycine} + 2H_2O \rightarrow 3\ \text{acetic acid} + CO_2 + 3NH_3$$

Single amino acids can also be fermented. Thus, certain clostridia ferment alanine according to the following equation:

$$3\ \text{alanine} + 2H_2O \rightarrow 2\ \text{propionic acid} + \text{acetic acid} + CO_2 + 3NH_3$$

In this fermentation one alanine is oxidized to acetic acid, ammonia and carbon dioxide as in the first example. The other two alanines are reduced as follows:

$$2CH_3\text{–}\underset{\underset{\text{alanine}}{NH_2}}{\underset{|}{CH}}\text{–}COOH + 4H \rightarrow \underset{\text{propionic acid}}{2CH_3CH_2COOH} + 2NH_3$$

Much less is known about these fermentations than about the fermentations of glucose. In particular, it is not known where or how much ATP is produced. However it clearly is produced, since the clostridia can grow using these fermentations as the only source of energy.

RESPIRATION

Respiration is an energy-yielding oxidation–reduction reaction in which the oxidant is inorganic. In the course of their evolution, bacteria have developed a wide variety of different kinds of respiration which can be characterized on the basis of the nature of the reductant and of the oxidant. Table 4-2 shows the oxidants and reductants in the various bacterial respirations.

Aerobic Respiration with Organic Reductants

The discussion of bacterial respiration will begin with the kind most familiar to you: the oxidation of organic compounds by oxygen or, more

TABLE 4-2

Reductants and Oxidants in Bacterial Respirations

Reductant	Oxidant	Products	Organism
H_2	O_2	H_2O	hydrogen bacteria
H_2	SO_4^{2-}	$H_2O + S^{2-}$	*Desulfovibrio*
organic compounds	O_2	$CO_2 + H_2O$	many bacteria, all plants and animals
NH_3	O_2	$NO_2^- + H_2O$	nitrifying bacteria
NO_2^-	O_2	$NO_3^- + H_2O$	nitrifying bacteria
organic compounds	NO_3^-	$N_2 + CO_2$	denitrifying bacteria
Fe^{2+}	O_2	Fe^{3+}	*Ferrobacillus*
S^{2-}	O_2	$SO_4^{2-} + H_2O$	*Thiobacillus*

briefly, aerobic respiration. In *Cell Structure and Function* the main features of the cellular aerobic respiration of higher organisms were introduced. The respiration of organic substrates by bacteria is, in most respects, very similar. Cellular respiration comprises two different phases. The first includes reactions by which the substrate is oxidized to CO_2 with the successive removal of pairs of hydrogen atoms. The series of reactions known as the *Krebs cycle* plays an indispensable role in this phase. The second phase is the oxidation of these hydrogen atoms by oxygen with the formation of ATP. Together, the two phases lead to oxidation of the substrate to CO_2 and water, and to the formation of biologically useful energy (ATP).

The Krebs cycle (see Fig. 4-3) is of central importance in bacterial respiration. It was noted previously that in their role as scavengers, the bacteria as a group must be able to oxidize every organic compound found in living organisms. The metabolism of each of these compounds involves a series of reactions, each catalyzed by a specific set of enzymes, leading eventually to a Krebs cycle intermediate, which is oxidized to CO_2 and water by the further reactions of the cycle.

A given type of bacterium can oxidize only a limited number of compounds. Some bacteria can attack a very broad range of substrates, whereas others are much more restricted. In every instance, however, the Krebs cycle plays the same central role.

All these reactions themselves contribute little in the way of energy to the cell. Most of the ATP is produced during the transfer of hydrogen atoms from the substrate to oxygen. The hydrogen atoms come from the dehydrogenations of the Krebs cycle; and, in addition, the reaction sequences by which Krebs cycle intermediates are formed from different substrates usually include dehydrogenations. Regardless of their source, all hydrogen atoms are handled in the same way. In the cells of higher animals, the transport is mediated by the cytochrome pigments. These pigments form a series or a ladder of oxidation–reduction reactions. The cytochrome at the bottom of the ladder is reduced by the reduced DPN formed in dehydrogenations. This first cytochrome is reoxidized by the next cytochrome, which in turn reduces the next, being itself reoxidized, and so on. The last cytochrome on the ladder is oxidized by oxygen, and water is produced. The energy of the reaction between hydrogen and oxygen is used to form ATP. Three molecules of ATP are formed for every pair of hydrogen atoms oxidized.[1]

The situation is much the same in bacterial respiration. Unfortunately, knowledge of the bacterial cyctochromes is insufficient to permit drawing a detailed picture. In any event, it is certain that ATP formation is coupled with the oxidation of hydrogen.

[1] The oxidation of succinic acid to fumaric acid is an exception; here only two ATP's are produced.

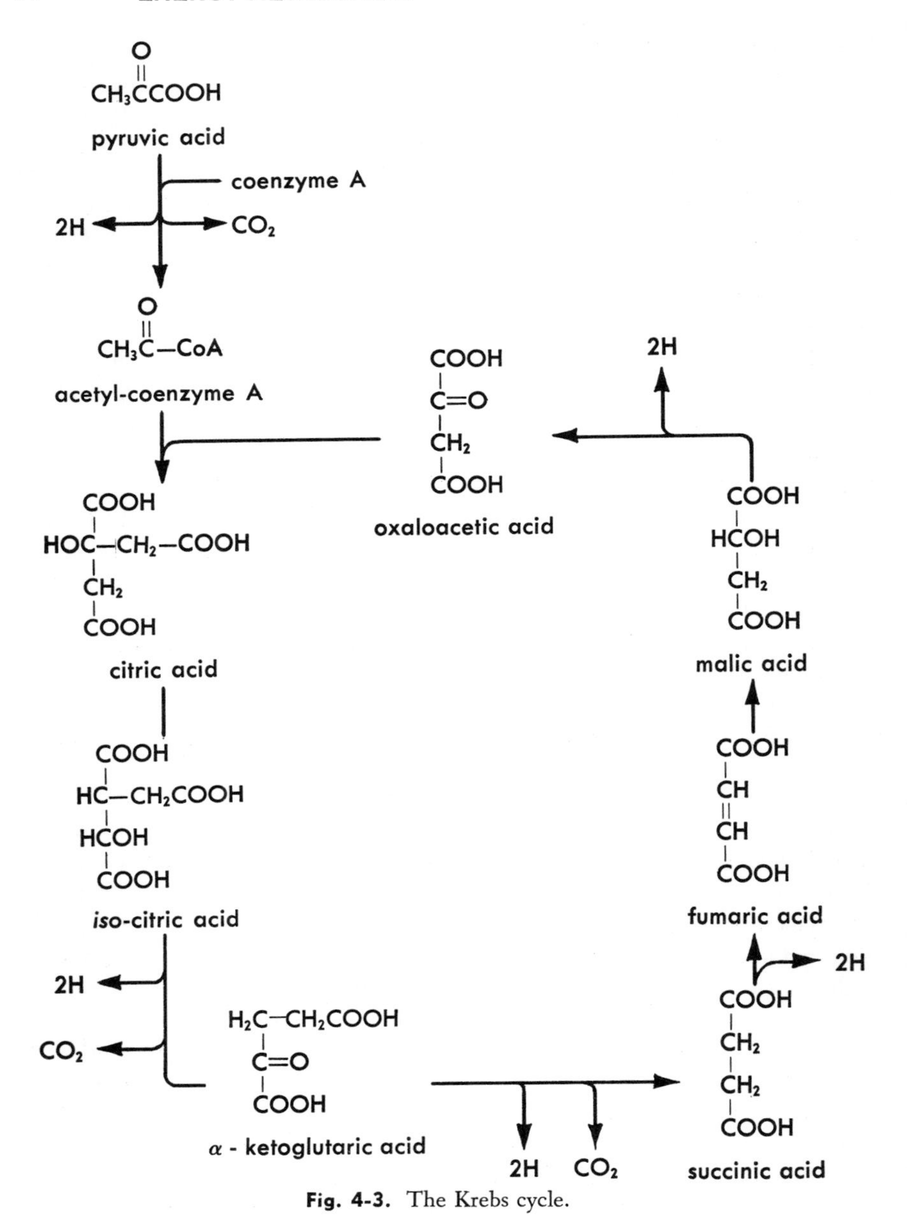

Fig. 4-3. The Krebs cycle.

Anaerobic Respiration

Up to now, bacterial respiration in which oxygen is the oxidant has been considered. There is another type of respiration in which oxygen is replaced by some other inorganic oxidant.

Bacteria of the genus *Desulfovibrio* oxidize organic compounds using sulfate as the oxidant, the sulfate being reduced to sulfide. The sulfate reducers cannot use oxygen.

Nitrate is used by many bacteria as the oxidant in respiration. The denitrifying bacteria, as they are called, can also use oxygen as the oxidant. These bacteria reduce nitrate only in the absence of oxygen. The nitrate is reduced to nitrogen gas or to ammonia or nitrous oxide, depending on the bacterium.

In both examples of anaerobic respiration, cytochromes play their usual role in hydrogen transport. Thus, the only difference between aerobic and anaerobic respiration is the nature of the final oxidant. Fig. 4-4 is an outline of the major features of bacterial respiration.

Incomplete Oxidations

Respiration usually results in the complete oxidation of the organic substrate. In other words, the only end products are CO_2, water, and cell material. In some bacteria and in many fungi, the oxidation is incomplete and organic end products accumulate. The respiration of these organisms is sometimes erroneously spoken of as "aerobic" or "oxidative" fermentation. They have no more in common with fermentation than does the more usual complete respiration. Although organic compounds are end products, they are always oxidized with respect to the substrate; in fermentation, however, both oxidized and reduced end products are formed. Furthermore, incomplete oxidations require the presence of oxygen, whereas fermentation is a nonoxygen-requiring or anaerobic process.

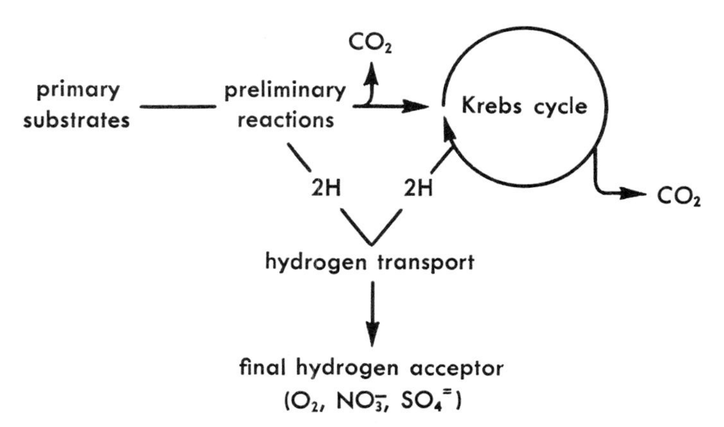

Fig. 4-4. Salient features of bacterial respirations. The "primary substrates" include all the many compounds attacked by bacteria. The "final hydrogen acceptor" is oxygen in aerobic respiration and nitrate or sulfate in anaerobic respiration.

Several of the incomplete oxidations are of considerable economic importance. For example, vinegar is produced from wine (alcohol) by the action of the acetic acid bacteria, which oxidize ethanol to acetic acid. The same bacteria also carry out incomplete oxidations of glucose to gluconic acid and ketogluconic acids.

Fungi are notable for the great variety of products they produce by the incomplete oxidation of sugars. A striking example is furnished by *Aspergillus niger* which converts almost 60 percent of the glucose it metabolizes to citric acid. This oxidation was for a long time the principal source of citric acid.

It is to be emphasized that organisms which carry out incomplete oxidations do not lack completely the metabolic systems necessary for the complete oxidation of the substrate; for example, *Aspergillus niger* can oxidize citric acid and can indeed grow by a respiration with citric acid as the substrate. Apparently, this fungus converts glucose to citric acid much more rapidly than it can oxidize citric acid, which therefore accumulates. This same mechanism is undoubtedly the basis of other incomplete oxidations.

Respiration with Inorganic Reductants

It was stated earlier (p. 36) that bacteria could be divided into organotrophs and lithotrophs. The energy metabolism of organotrophs is based on the oxidation of organic compounds and that of lithotrophic bacteria on the oxidation of inorganic compounds. Compounds of hydrogen, iron, ammonia, and nitrate are known to be oxidized by these bacteria, as are a variety of sulfur compounds. The lithotrophic bacteria are further distinguished from organotrophs in their ability to derive all their cellular carbon from CO_2. Organisms which can grow in the absence of any organic carbon are called autotrophs. Lithotrophic bacteria were the first example of autotrophy discovered outside of the green plants. It should be noted that autotrophy is not *necessarily* associated with a lithotrophic energy metabolism. Although all lithotrophs so far discovered are autotrophs, the existence of bacteria that require organic carbon but obtain their energy from the respiration of inorganic substrates seems possible.

Oxidation of hydrogen

The oxidation of gaseous hydrogen is used by a group of bacteria known as hydrogen bacteria. Their respiration is represented by the following equation:

$$H_2 + \tfrac{1}{2}O_2 \rightarrow H_2O + \text{energy}$$

In contrast to other lithotrophic bacteria, hydrogen bacteria are also capable of respiring organic compounds.

The key reaction in the oxidation of hydrogen is the activation of hydrogen. In this reaction a molecule of hydrogen is split to give two hydrogen atoms:

$$H_2 \rightleftharpoons 2H$$

This reaction, not peculiar to the hydrogen bacteria, is found in many organotrophic bacteria. The oxidation of the hydrogen atoms so formed is essentially the same as in the respiration of organic compounds.

The nitrifying bacteria

The oxidation of ammonia to nitrate is an important step in the economy of nitrogen in nature. This nitrification process occurs in two steps, each carried out by a very specialized group of bacteria. The first step is the oxidation of ammonia to nitrite.

$$NH_3 + 1\tfrac{1}{2}O_2 \rightarrow NO_2^- + H^+ + H_2O + \text{energy}$$

Bacteria of the genus *Nitrosomonas* can use this oxidation as their sole source of energy; CO_2 serves as carbon source. These bacteria are incapable of any other mode of growth.

The second step in nitrification is the oxidation of nitrite to nitrate.

$$NO_2^- + \tfrac{1}{2}O_2 \rightarrow NO_3^- + \text{energy}$$

The important agents here are included in the genus *Nitrobacter;* the oxidation of nitrite is the only source of energy these bacteria can use.

Sulfur bacteria

The oxidation of reduced sulfur compounds is the energy source for several bacteria. One of these, *Thiobacillus thiooxidans,* oxidizes elemental sulfur to sulfuric acid:

$$S^0 + H_2O + 1\tfrac{1}{2}O_2 \rightarrow H_2SO_4 + \text{energy}$$

This organism is unique in its ability to grow in extremely acid environments, as indeed it must be if it is to make use of the formation of sulfuric acid as an energy source.

Another group of sulfur oxidizing bacteria are morphologically closely related to the blue-green algae. This group includes *Beggiatoa,* which oxidizes hydrogen sulfide to elemental sulfur.

$$H_2S + \tfrac{1}{2}O_2 \rightarrow S^0 + H_2O$$

The long filaments of *Beggiatoa* become stuffed with minute granules of sulfur, which give it a very characteristic appearance under the microscope.

Iron bacteria

The final example of lithotrophic metabolism is perhaps the most interesting of all; this is the oxidation of iron carried out by *Ferrobacillus.* The energy-yielding oxidation is simply:

$$Fe^{2+} \rightarrow Fe^{3+} + e^-$$

The ferric iron is deposited as insoluble ferric hydroxide. Several other bacteria oxidize iron with the formation of ferric hydroxide, but whether these can use the oxidation as a source of energy is not certain.

Not a great deal is known about hydrogen transport and ATP formation in lithotrophs. However, it is very probable that here, too, cytochrome pigments play a central role.

THE INFLUENCE OF OXYGEN ON THE GROWTH OF BACTERIA

The type of energy metabolism which a bacterium possesses determines to a large extent the environment in which it can live. From the point of view of energy metabolism, the most important single factor in the environment is the presence or absence of oxygen. In terms of their response to the presence of oxygen, bacteria fall into three groups: obligate aerobes, obligate anaerobes, and facultative anaerobes.

The obligate aerobes can grow only in the presence of oxygen. Oxygen is necessary for these bacteria because aerobic respiration is the only kind of energy metabolism they possess. For the same reason, most animals are also obligate aerobes.

Obligate anaerobes can grow only in the absence of oxygen. The growth of these organisms is inhibited by oxygen; indeed, some of them are so sensitive to the presence of oxygen that a brief exposure to air kills them. The basis for this sensitivity to oxygen is not entirely clear; but it can be explained, in part at least, by the destruction of certain indispensable enzymes by oxidation. It is obvious that obligate anaerobes must be able to obtain energy from processes which do not require oxygen, such as fermentation or anaerobic respiration.

The facultative anaerobes can grow either in the presence or in the absence of oxygen. The facultative anaerobes include three kinds of bacteria with very different physiologies. The first kind includes bacteria which have a purely fermentative energy metabolism, and thus are indifferent to oxygen. The second kind of facultative anaerobes are the denitrifying bacteria. The third and largest kind include bacteria and fungi that obtain the energy for growth from either fermentation or aerobic respiration. Ordinary baker's yeast is an excellent example of this kind of organism.

It is instructive to contrast the aerobic and anaerobic metabolism of the last two kinds of facultative anaerobes. In the first place, the two have different requirements for anaerobic growth. The denitrifying bacteria can use, anaerobically, any substrate they can use aerobically, provided that nitrate is also present. In contrast, organisms that can ferment do not need a substitute for oxygen in order to grow anaerobically; but, on the other hand, not all substrates of aerobic growth are fermentable. Thus, succinic acid can be respired by many such organisms, but it cannot be fermented; consequently, it will not support anaerobic growth. A fermentable substrate, such as glucose, can be used with equal facility aerobically and anaerobically.

A second and more important difference is that a denitrifying bacterium respires whether it is growing aerobically or anaerobically, whereas the other kind of facultative anaerobes ferment under anaerobic conditions and respire under aerobic conditions. In the denitrifying bacteria only the nature of the final hydrogen acceptor changes; in the second case the energy metabolism as a whole changes. The change in the kind of energy metabolism is more than simply the addition of respiration; it involves the suppression of fermentation by oxygen as well.

This remarkable effect of oxygen was first noted by Pasteur while comparing the growth of yeast under aerobic and anaerobic conditions. Pasteur also observed that more yeast is formed from a given amount of glucose aerobically than anaerobically; or in other words, that respiration is more efficient than fermentation. Both these fundamental observations of Pasteur are undoubtedly reflections of the different yields of ATP in fermentation and respiration.

It has been shown that during glycolysis the formation of two molecules of pyruvic acid is accompanied by the net formation of two molecules of ATP. Respiration of glucose yields much more ATP. In the first place all of the hydrogen atoms are oxidized, and in the second place the transfer of a pair of hydrogen atoms to oxygen provides more energy than the transfer to an organic compound.

The amount of energy used in the formation of a certain amount of cellular material does not depend upon how the energy is obtained. Therefore, as more ATP is produced from a given amount of substrate, more cell material is formed. It can be seen then why respiration of a given amount of a substrate gives more cell material than does fermentation of the same amount. Some examples of the relation between ATP yield and amount of growth are shown in Table 4-3.

This table shows the amount of cell material formed in terms of dry weight per mole (180 grams) of glucose fermented or respired. In the fermentations shown, all the ATP is formed in the conversion of glucose to pyruvic acid; thus, the amount of ATP formed is known exactly. Two glycolytic fermentations are included. In the yeast fermentation, the products are ethanol and CO_2; in the lactic acid bacterium, lactic acid is the only product. Both give exactly the same amount of growth, 20 grams per mole of glucose. The amount of cell material formed for every ATP produced is easily calculated to be 10 grams. The fermentation by *Zymomonas* (a bacterium) gives the same products as the yeast fermentation, but the fermentation involves the Entner-Doudoroff pathway instead of glycolysis and, accordingly, only one ATP is formed. Only one half as much cell material is formed; the cell material formed per ATP, however, is the same. The HMP fermentation is characteristic of another kind of lactic acid bacterium. The products, in addition to lactic acid, include ethanol and CO_2. It was shown previously that this pathway, like the Entner-Doudoroff pathway, yields only one ATP; and the table shows that the amount of growth is the same as in the case of *Zymomonas*. Finally, the result with respiring yeast shows the magnitude of the difference between respiration and fermentation; that is, in terms of glucose used, respiration is five times as efficient as fermentation.

The other effect noted by Pasteur—the inhibition of fermentation by respiration—is undoubtedly related to the higher yield of ATP from respiration. But until much more is known about how a cell controls its own metabolic processes, the mechanism of the effect must remain obscure.

BACTERIAL PHOTOSYNTHESIS

The second source of energy utilized by living things is sunlight. Organisms, such as green plants and algae, that make use of this source of energy are called phototrophs. From what you have seen so far of the enormous diversity of metabolism among the bacteria, it will perhaps not

TABLE 4-3

Growth Yields in Fermentation and Respiration of Glucose

Organism	Kind of energy metabolism	Grams dry weight per mole glucose	ATP per mole glucose	Dry weight per ATP
yeast	glycolysis	20	2	10
yeast	respiration	about 100	high	?
lactic acid bacterium	glycolysis	20	2	10
lactic acid bacterium	HMP fermentation	10	1	10
Zymomonas	ED fermentation	10	1	10

come as a surprise to learn that there is a group of phototrophic bacteria. These bacteria are of little importance ecologically compared to the green plants and algae. However, the study of bacterial photometabolism has made signal contributions to our understanding of photosynthesis in general.

Since photosynthesis is, ultimately, the transformation of radiant energy into chemical energy, it is necessary to digress briefly and consider some simple facts concerning light as energy. Light may be considered as consisting of discrete particles called photons. Each photon has a certain amount of energy associated with it. The amount of this energy depends upon the wavelength of the light; the shorter the wavelength, the greater the energy content of the photon. Thus, for example, a photon of blue light has more energy associated with it than one of red light.

Light energy can be converted to chemical energy only if absorbed. Compounds which strongly absorb light are called pigments. Pigments are colored because they absorb light of only certain wavelengths, that is, they absorb photons of certain energies. A pigment molecule either absorbs a photon or it does not; it cannot absorb a part of a photon. The energy of the absorbed photon is transferred to the pigment molecule. Since the molecule now has more energy than it normally has, it is said to be excited. An excited molecule can get rid of this excess energy and return to its original or ground state in a number of ways. The way of interest here involves the use of the energy to drive a chemical reaction. In essence, this is all there is to photosynthesis. What distinguishes phototrophs is not the ability to convert radiant energy into chemical energy—you do this whenever you get a sunburn—but to convert it into useful chemical energy.

The Kinds of Photosynthetic Bacteria

Simply on the basis of color, the photosynthetic bacteria fall into two principal groups: the green and the purple bacteria. As far as their morphology is concerned, all are typical bacteria.[1] They are gram-negative rods or spirilla, when motile the flagella are polar, and no endospores are produced.

The photosynthetic bacteria, like green plants, contain chlorophylls. The chlorophyll found in the green bacteria is chemically very similar to the chlorophyll of higher plants and is responsible for the characteristic color of these bacteria.

The chlorophyll found in the purple bacteria (bacteriochlorophyll) differs chemically from green plant chlorophyll in several ways. One result of this is that bacteriochlorophyll is not green but is a pale blue-gray. The color of the purple bacteria is caused by the presence of various carotenoid pigments

[1] With the exception of *Rhodomicrobium,* a nonsulfur purple bacterium, which is one of the budding bacteria mentioned on page 15.

which in these bacteria are yellow and red. (The carotenoid pigments of the green bacteria are yellow, and hence do not mask the color of chlorophyll.)

The carotenoid pigments are accessory pigments; this means that the light energy absorbed by them can be used in photosynthesis but only be being transferred to chlorophyll. This property is of considerable advantage to the organism, since light not absorbed by chlorophyll itself is still available for photosynthesis.

The Physiology of the Photosynthetic Bacteria

The purple bacteria can be divided into two kinds: the sulfur purple and the nonsulfur purple bacteria. The essential features of the physiology of the green bacteria and of the sulfur purple bacteria are very similar and will be considered first. The nonsulfur purple bacteria have a different physiology, which is more easily understood after the principal facts about bacterial photosynthesis have been introduced.

Both the sulfur purple and the green bacteria will grow if provided with light, CO_2, and one of a number of inorganic sulfur compounds, such as hydrogen sulfide. Analysis of a culture before and after growth shows that the sulfide has been oxidized to sulfate and that the CO_2 has been transformed into cell material. The process can be represented by the following equation:

$$2H_2O + H_2S + 2CO_2 \xrightarrow{\text{light}} 2(CH_2O) + SO_4^{2-} + 2H^+$$

Here (CH_2O) represents cell material. A more careful analysis reveals that this transformation occurs in two stages which can be represented by the following equations:

$$\text{(a)}\quad H_2S + \tfrac{1}{2}CO_2 \xrightarrow{\text{light}} S^0 + \tfrac{1}{2}(CH_2O) + \tfrac{1}{2}H_2O$$

$$\text{(b)}\quad S^0 + 2\tfrac{1}{2}H_2O + 1\tfrac{1}{2}CO_2 \xrightarrow{\text{light}} SO_4^{2-} + 1\tfrac{1}{2}(CH_2O) + 2H^+$$

The first stage (a) predominates so long as hydrogen sulfide is available in the medium; the sulfur is usually deposited intracellularly, but with some smaller bacteria is found in the medium.

The similarity of this metabolism to that of the nonphotosynthetic sulfur bacteria should be noted (see p. 47).

It is instructive to compare these two metabolisms in more detail. The chemotrophic sulfur bacteria obtain energy from the oxidation of sulfide (or other sulfur compounds); the hydrogen acceptor is oxygen. This energy, in the form of ATP, is used for growth. These bacteria, being autotrophs, use CO_2 as carbon source. The reduction of CO_2 to cell material (CH_2O)

requires energy and some of the ATP is used for this purpose. The reduction of CO_2 also requires a source of hydrogen. Ultimately, this hydrogen is supplied by the reduced sulfur compound. In other words, some of the substrate is used to reduce oxygen (producing H_2O and ATP) and some is used to reduce CO_2 (producing cell material). The total amount of substrate oxidized is thus equal to that necessary to supply hydrogen for the reduction of CO_2 plus that needed to supply the energy for growth.

The sulfur purple and the green bacteria are also autotrophs and so require energy and hydrogen to reduce CO_2. As in the chemotrophic sulfur bacteria, the hydrogen comes from the reduced sulfur compound. However, in the phototrophs, CO_2 is the only compound present that can accept hydrogen. The amount of sulfide oxidized is, therefore, equivalent to the CO_2 reduced, as can be seen from the equation. How then does the phototroph acquire energy for growth? Obviously from light. The only thing unique to phototrophs is the ability to convert light energy into chemical energy in the form of ATP. The ability to reduce CO_2 is shared with the colorless sulfur bacteria and with all chemotrophic autotrophs. Indeed, as shown in the next chapter, both kinds of autotrophs use the same enzymic reactions to accomplish this reduction.

In the metabolism of the nonsulfur purple bacteria some organic compound replaces sulfur. Light is still the source of ATP. These bacteria stand in the same relation to the sulfur purple and green bacteria as do organotrophic bacteria to the colorless sulfur bacteria. The nonsulfur purple bacteria can and do reduce CO_2, but the extent of this reduction depends upon the organic substrate. When the substrate is more reduced than cell material (CH_2O)—that is, when the ratio of hydrogen to oxygen is greater than two to one—the excess hydrogen is used to reduce CO_2. If, on the other hand, the substrate is more oxidized, CO_2 is given off just as in a respiration. When, for example, propionic acid is the substrate, CO_2 is reduced. The empirical formula of propionic acid is $C_3H_6O_2$, hence, when it is converted to cell material, CO_2 must be reduced.

$$2C_3H_6O_2 + H_2O + CO_2 \rightarrow 7(CH_2O)$$

However, if malic acid ($C_4H_6O_5$) is the substrate, CO_2 is evolved.

$$C_4H_6O_5 \rightarrow 3(CH_2O) + CO_2$$

All the green and sulfur purple bacteria are obligate anaerobes; on the other hand, several representatives of the nonsulfur purple bacteria can grow in the presence of oxygen. These bacteria not only are not inhibited by oxygen, but also can carry out aerobic respiration; thus they are able to grow in the dark if provided with oxygen. None of the photosynthetic bacteria can grow anaerobically in the dark, since they can neither ferment nor carry out an anaerobic respiration.

Photosynthetic Bacteria and Green Plants

We have seen that bacterial photosynthesis can be described as an oxidation–reduction reaction in which CO_2 is the oxidant and either an inorganic sulfur compound or an organic compound is the reductant. The energy for this reaction is provided by light. The overall process of bacterial photosynthesis can be represented by the following equation:

$$2H_2A + CO_2 \xrightarrow{\text{light}} 2A + (CH_2O) + H_2O$$

The H_2O comes from the reduction of CO_2

$$4H + CO_2 \rightarrow (CH_2O) + H_2O$$

These hydrogen atoms are provided by the substrate, H_2A

$$2H_2A \rightarrow 2A + 4H$$

In the case of the sulfur bacteria, H_2A is hydrogen sulfide (H_2S) and A is sulfur.

The classic equation for green plant photosynthesis is

$$CO_2 + H_2O \xrightarrow{\text{light}} (CH_2O) + O_2$$

The equation is simply a description of the results of analysis of the substances disappearing and appearing during photosynthesis. As it stands, this equation cannot be written as the sum of an oxidation and a reduction. (The reader is urged to try.) However, if the equation is modified by the addition of water to each side:

$$CO_2 + 2H_2O \xrightarrow{\text{light}} (CH_2O) + H_2O + O_2$$

it becomes obvious what the oxidation and reduction reactions are. The oxidation is:

$$2H_2O \rightarrow 4H + O_2$$

and the reduction is:

$$4H + CO_2 \longrightarrow (CH_2O) + H_2O$$

Thus the two kinds of photosynthesis, bacterial and green plant, can be represented by the same general equation:

$$2H_2A + CO_2 \xrightarrow{\text{light}} (CH_2O) + 2A + H_2O$$

In green plants H_2A is H_2O and $2A$ is O_2; in bacteria H_2A is some oxidizable substrate and $2A$ is the product of its oxidation. It is perhaps worthwhile remarking why an examination of bacterial photosynthesis revealed so clearly that the process is fundamentally an oxidation–reduction reaction. It is simply that with bacteria, one of the products of the reduction of CO_2, namely water, is not the same as the oxidizable component, hydrogen sulfide for example. In green plants, both are water and hence the product of the reduction could not be distinguished from the oxidant by simple chemical analysis.

The Role of Light

The unique feature of photosynthesis is the conversion of light energy to the chemical energy of ATP. A detailed discussion of the biochemical mechanism of this conversion can be found in *The Living Plant,* in this series. Here we can only review those parts of the process that are pertinent to an understanding of bacterial photosynthesis.

The energy absorbed by chlorophyll is used to form an oxidant and a reductant, and thus represents the essential conversion of light energy into chemical energy. Just how the excited chlorophyll does this and the identities of the oxidant and the reductant are not known with complete certainty. In general the reactions can be written as follows:

$$\text{chlorophyll} + \text{light} \rightarrow \text{chlorophyll}^* \text{ (excited)}$$

$$\text{chlorophyll}^* + A_R + B_O \rightarrow \text{chlorophyll} + A_O + B_R$$

where A and B represent the oxidant and the reductant formed from light energy. In other words, the reaction

$$A_O + B_R \rightarrow A_R + B_O$$

is exergonic, and the energy provided by excited chlorophyll is required to drive it in the opposite direction. Notice that either A or B could be identical to chlorophyll. If for example A were chlorophyll, the reaction would be:

$$\text{chlorophyll}^* + B_O \rightarrow \text{chlorophyll}_O + B_R$$

But the precise nature of A or B is really a detail. The oxidant (A_O) and reductant (B_R) can now react and the energy available from this reaction can be used to form ATP. The way in which this occurs appears to be in all essentials the same as in respiration. Electrons are transported from B_R to A_O via a chain of cytochromes and ATP formation is coupled with the oxidation and reduction of cytochromes. We can write these reactions as follows:

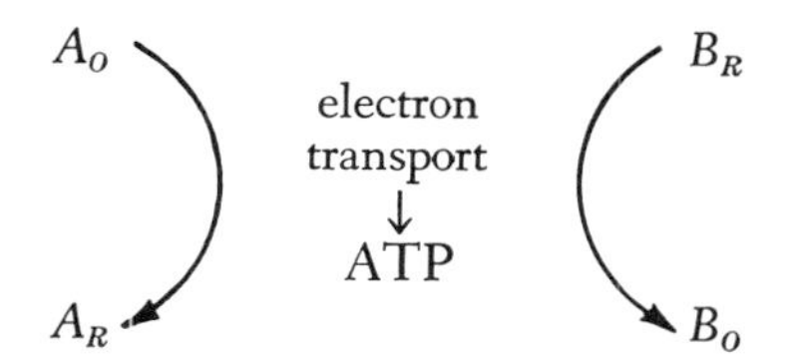

The whole process is, then, in essence:

$$\text{light} \rightarrow \left(A_O + B_R \;\rightleftarrows\; A_R + B_O \right) \rightarrow \text{ATP}$$

In contrast to respiration neither the oxidant nor the reductant in the ATP-generation process is used up. The photochemically produced reductant is used only if CO_2 reduction occurs. When this happens there is an equivalent amount of oxidant remaining. In the bacteria this oxidant is reduced by the oxidizable substrate (H_2S, or organic compounds); in green plants it appears as O_2.

To summarize, in both respiration and in photosynthesis ATP is formed in essentially the same way: by the transfer of electrons from a reductant to an oxidant via cytochromes. These two kinds of metabolism differ in the source of the reductant and the oxidant. In respiration these are provided in the environment, and are ready-made, so to speak. In photosynthesis these are formed within the cell at the expense of light energy.

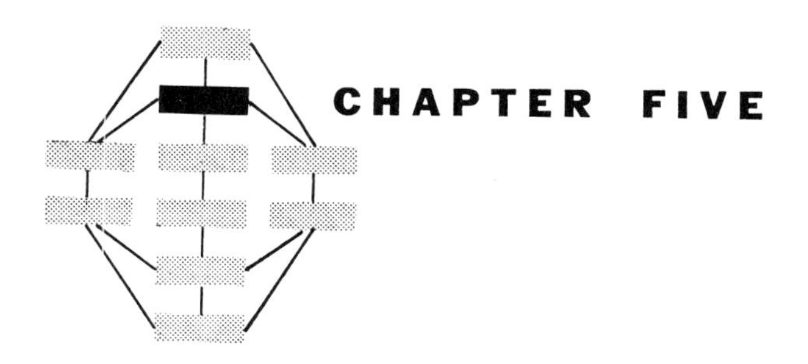

CHAPTER FIVE

BACTERIAL NUTRITION AND THE ECOLOGY OF BACTERIA

The chemical composition of the bacterial cell was discussed in Chapter 3. There are two sorts of cell components: small molecules and macromolecules. The macromolecular components include the proteins, nucleic acids, and polysaccharides. The specificity of the cell resides in these components; that is, one bacterium differs from another because, to a greater or lesser extent, the macromolecules are different. The low molecular weight components are, on the contrary, identical in all bacteria and indeed in all organisms. Included here are the coenzymes (ATP, DPN, TPN, coenzyme A, etc.) and also the prosthetic groups of certain enzymes. These molecules are important to the cell in themselves. A second group of small molecules includes the building blocks of macromolecular components. These are important only because of what they will become.

The previous chapter described the ways in which bacteria obtain energy. They use this energy for acquiring small molecules and for synthesizing their macromolecular components. This chapter will investigate the ways bacteria obtain the small molecules needed for growth. The question of macromolecular synthesis is deferred until the next chapter.

THE SPECTRUM OF NUTRITIONAL TYPES

All organisms grow by assimilating material from their environment and all the substance of an organism is derived from its environment. This statement, which is a biological rendering of the Law of Conservation of Matter, is so obvious that it sometimes escapes attention, but it leads us to go on to ask: what kinds of material do different organisms require from the environment or, in other words, what are the nutritional requirements of different organisms? Even a superficial survey reveals that there is an immense range of nutritional requirements. At one extreme are the green plants, with very simple requirements, and at the other extreme are animals, with complex needs.

Green plants need only sunlight, carbon dioxide, water, and some inorganic nitrogen in order to grow. On the other hand, animals must feed on other organisms. Green plants are called autotrophs and animals are called heterotrophs. We have already encountered several autotrophic bacteria. As we shall see, there are also bacteria with nutritional requirements as complex as those of an animal.

The nutritional requirements of a bacterium (or any organism) make manifest its biosynthetic capabilities. There are only two ways for a cell to obtain the molecules it needs in order to grow. Either they can be provided ready-made by the environment, or they can be synthesized by the cell from other materials from the environment. Since the molecular compositions of all bacteria are the same, it is clear that the nutritional requirements and biosynthetic capabilities of a cell are inversely related to each other.

Autotrophic bacteria can synthesize all of their own small molecules. A vast number of organotrophic bacteria (those which use organic compounds as energy sources) can grow when supplied with a single organic compound. The biosynthetic abilities of these bacteria are fully as great as those of the autotrophs. The ability of these bacteria to make all their own building blocks means that they can form the enzymes which catalyze a vast network of biosynthetic reactions. A bacterium which lacks one or more of these enzymes cannot synthesize certain building blocks, and so these become required nutrients.

In this chapter some of the major features of this network are examined. The details of the individual reaction sequences are omitted.

Autotrophic Fixation of CO_2

Autotrophs acquire the bulk of their carbon from carbon dioxide. The mechanism of fixation of CO_2 appears to be identical in all autotrophs—green plants, photosynthetic bacteria, and chemolithotrophic bacteria. The key reaction is the addition of CO_2 to ribulose-diphosphate, forming phosphoglyceric acid.

$$\begin{array}{ll} CH_2OPO_3H_2 & \\ \vert & \\ C{=}O & CH_2OPO_3H_2 \\ \vert & \vert \\ CHOH + H_2O + CO_2 \rightarrow & 2CHOH \\ \vert & \vert \\ CHOH & COOH \\ \vert & \\ CH_2OPO_3H_2 & \\ \text{ribulose-diphosphate} & \text{phosphoglyceric acid} \end{array}$$

The phosphoglyceric acid can be metabolized in at least two ways. In the first place it can be reduced to glyceraldehyde phosphate. This is the reverse of the oxidation of the latter compound which occurs during glycolysis (see *Cell Structure and Function*):

$$\begin{array}{ll} CH_2OPO_3H_2 & CH_2OPO_3H_2 \\ \vert & \vert \\ CHOH + ATP + 2H \rightarrow & CHOH + ADP \\ \vert & \vert \\ COOH & CHO \\ \text{phosphoglyceric acid} & \text{glyceraldehyde phosphate} \end{array}$$

The hydrogen atoms are supplied by either diphosphopyridine nucleotide or the functionally similar triphosphopyridine nucleotide. The source of the ATP is the only real difference between phototrophs and other autotrophs.

Alternatively, phosphoglyceric acid can be converted to pyruvic acid by the same sequence of reactions as in glycolysis.

The substrate for the carboxylation reaction, ribulose-diphosphate, is formed by a series of reactions which was first demonstrated in green plants. The overall result is the conversion of five molecules of glyceraldehyde-phosphate to three molecules of ribulose-*mono*phosphate; the further conversion of these to ribulose-*di*phosphate requires three molecules of ATP. We have just seen that autotrophs can convert one molecule of ribulose-diphosphate and one molecule of CO_2 to two molecules of phosphoglyceric acid, and reduce this to glyceraldehydephosphate. In autotrophs, then, all these reactions form a cycle:

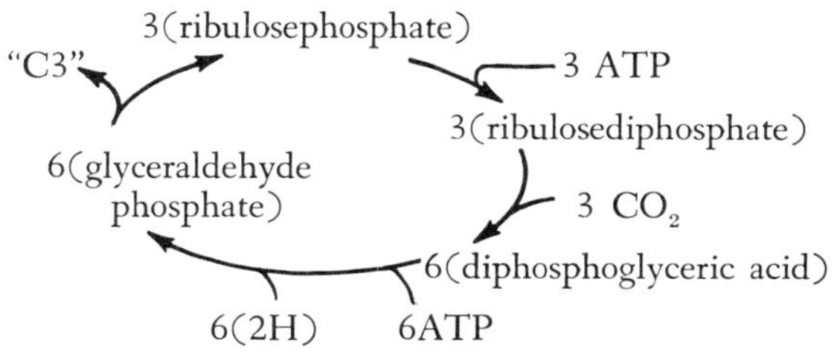

The C3 compound represents the net gain in carbon after one complete cycle. The details of the conversion of phosphoglyceric acid to ribulose phosphate are discussed in *The Living Plant* in this series; in green plants these reactions lead to the formation of starch. In photosynthetic bacteria, which do not form starch, the carbon from CO_2 is removed from the cycle at the phosphoglyceric acid stage and enters the Krebs cycle. Although the sequence of reactions which leads from phosphoglyceric acid to ribulose phosphate was first elucidated in green plants, it has subsequently been found to occur in many heterotrophic bacteria. In these organisms the cycle operates as an alternative mechanism of sugar metabolism especially when a 5-carbon sugar is metabolized.

The Biosynthesis of Building Blocks

Fig. 5-1 shows, in barest outline, the network of reactions by which the various small molecular components of a cell are made. The core of this network is made up of a small number of compounds or precursors. From these precursors, numerous reaction sequences lead to the building blocks. Many of these sequences branch, so that glutamic acid, for example, is itself a precursor of several other building blocks.

One of the most important facts in biology is that most of these precursors are intermediates of the Krebs cycle. The Krebs cycle is thus of central importance not only to the respiration of organic compounds but also to biosynthesis. It is present then in all bacteria, and indeed in all organisms, not just in those which respire organic substrates. This one central mechanism allows all carbon sources to be used to form the building blocks common to all bacteria. We have already seen that organic substrates of both fermentation and respiration are metabolized to Krebs cycle intermediates; the same is true of CO_2 in autotrophs.

There is one crucial difference between the Krebs cycle functioning in the oxidation of pyruvic acid and in biosynthesis. In the former case the intermediates are catalytic; they are not used up as the oxidation proceeds. In the latter case, the intermediates are not catalytic; instead, they are withdrawn as biosynthesis proceeds. No sooner are molecules of one of these intermediates formed than they begin to be transformed to building blocks and thence to become part of the fabric of the cell.

For the Krebs cycle to operate, there must be a constant supply of acetate (or something easily converted to acetate) and oxaloacetic acid. If, for example, a bacterium is growing on glucose, pyruvic acid will be constantly supplied; acetate is easily formed from pyruvic acid by any one of a number of different oxidations. The acetic acid (as acetyl coenzyme A) enters the Krebs cycle by reacting with oxaloacetic acid. As biosynthesis continues, the supply of the latter acid would dwindle and unless it were replenished the flow of carbon would become slower and slower. This is accomplished by the

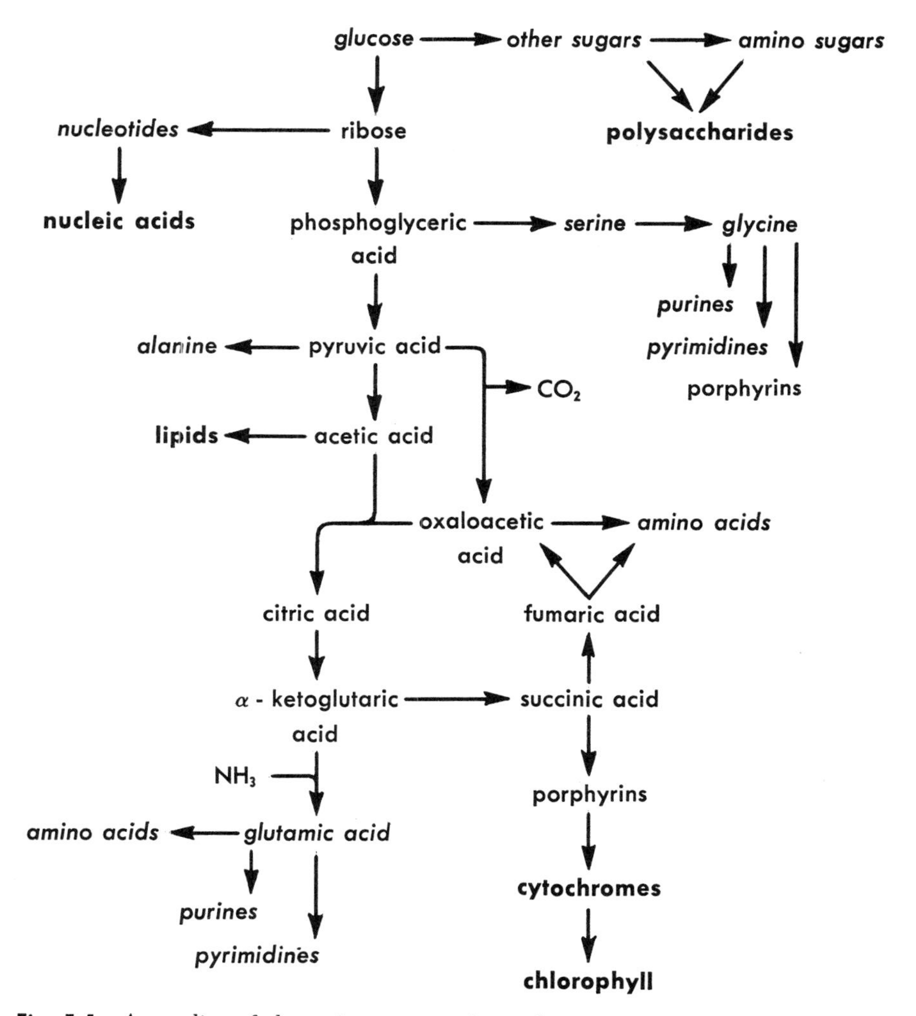

Fig. 5-1. An outline of the major routes in biosynthesis. Final products are shown in **bold face** type; building blocks in *italics*. Protein, the final product formed from the amino acids, is not shown.

synthesis of oxaloacetic acid from pyruvic acid and CO_2. This reaction is a complex one, but the overall result is represented by the following equation:

$$\underset{\text{pyruvic acid}}{CH_3COCOOH} + CO_2 + ATP \rightarrow \underset{\text{oxaloacetic acid}}{\begin{array}{l} COOH \\ | \\ C{=}O \\ | \\ CH_2 \\ | \\ COOH \end{array}} + ADP$$

This fixation of CO_2 occurs in all organisms—heterotrophs as well as autotrophs. It does not, of course, result in a net uptake of CO_2, since much more CO_2 is formed during respiration.

A bacterium using one of the intermediates of the Krebs cycle as a carbon source need not depend upon the reaction between pyruvic acid and CO_2 to maintain the supply of oxaloacetic acid.

A more serious problem is faced by the bacteria that use a two-carbon compound as sole source of carbon. The simplest solution would be to synthesize pyruvic acid from CO_2 and the two-carbon compound; however, such a C1-C2 condensation does not occur. Rather, a modified version of the Krebs cycle, known as the glyoxylic cycle, is used. The two key reactions in this cycle are (1) the splitting of *iso*-citric acid to succinic and glyoxylic acids and (2) the synthesis of malic acid from acetic acid (probably as acetyl coenzyme A) and glyoxylic acid. The cycle is shown in Fig. 5-2. The net result is the formation of one molecule of succinic acid (C4) from two molecules of acetic acid (C2).

$$2CH_3COOH \rightarrow COOHCH_2CH_2COOH + 2H$$

Growth factors and biosynthetic deficiencies

A bacterium in which part of the network of biosynthetic reactions is deficient must obtain from the environment the building blocks synthesized by that part of the network. A striking example is furnished by certain algae that, although the bulk of their carbon is derived from CO_2, must be furnished with very small amounts of a vitamin. Thus, a small section of the biosynthetic network is not functioning and the product must be supplied from the outside. Such specific requirements are called growth factors. In some cases the growth factor may be the same as the building block or it may be an intermediate in biosynthesis that the call can convert to the required product.

The greater the deficiencies in its biosynthetic machinery the more growth factors a bacterium needs. An extreme example is provided by some lactic acid bacteria. These bacteria obtain energy from the fermentation of glucose to lactic acid. Almost none of the glucose is used as a carbon source; 95 percent of the cell material comes from the specific growth factors which must be supplied. The medium for these bacteria contains a large portion of all of the known small molecular cell constituents.

These deficiencies in biosynthetic machinery are caused by the inability of the bacterium to form various enzymes. A requirement for a particular growth factor can always be correlated with the lack of one or more enzymes. These deficiencies are stable, heritable characters and are the result of mutations in the genetic material of the bacterium. The genetic basis of this is treated in detail in *Genetics,* in this series.

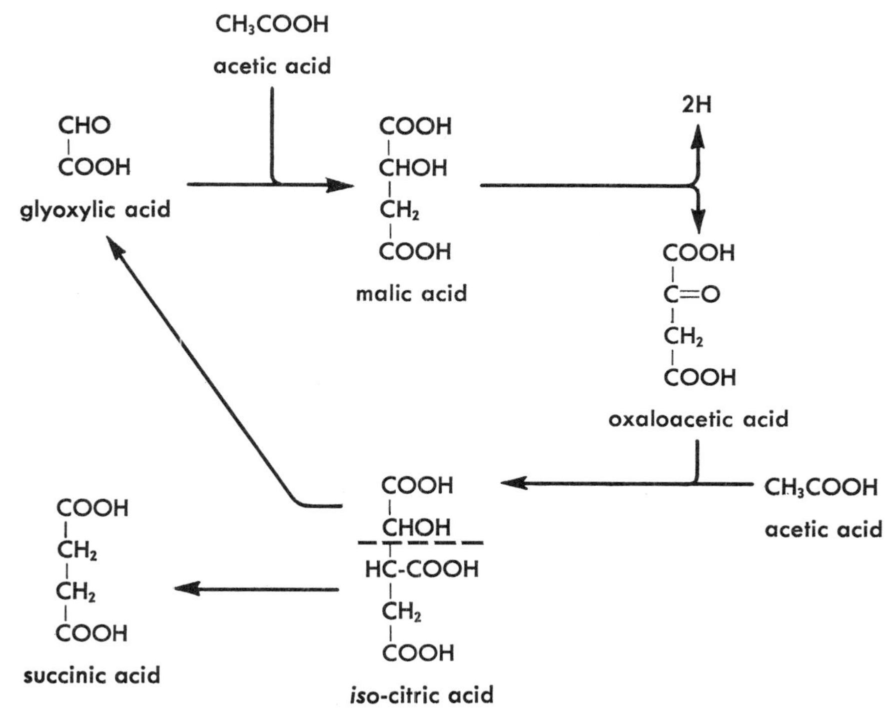

Fig. 5-2. The glyoxylic acid pathway. By making use of this cycle a bacterium can maintain a supply of the C_4 compounds needed in biosynthesis when using a C_2 compound, such as acetic acid, as the sole source of carbon.

BIOLOGICAL FIXATION OF NITROGEN

In autotrophic bacteria and in most heterotrophic bacteria, ammonia is used as the primary source of nitrogen. However, there is a group of microorganisms that can use atmospheric nitrogen as a nitrogen source. There are two types of such nitrogen-fixing bacteria: symbiotic and free-living.

The symbiotic nitrogen fixers infect the root hairs of certain kinds of plants, notably clover, alfalfa, and legumes. The plant responds to this infection by forming a small nodule within which the bacteria grow and multiply. Supplied with carbon compounds by the plant, the bacteria fix nitrogen producing ammonia and various amino acids. This process is the basis for the use of clover and alfalfa in crop rotation, since the growth of such plants increases the utilizable nitrogen in the soil. The symbiotic bacteria can be isolated and grown in pure culture; however, only the symbiont (the root nodule) can fix nitrogen.

The second group of nitrogen-fixing bacteria is larger and very heterogeneous; nitrogen fixation by these free-living forms is probably not of the same importance as the symbiotic nitrogen fixation. The two most important kinds of free-living nitrogen-fixing bacteria are *Azotobacter* and some *Clostridia*. *Azotobacter* are gram-negative, motile, short rods; they are rather large and frequently form gelatinous capsules. They are typical aerobic respiratory bacteria. The nitrogen-fixing *Clostridia* are strict anaerobes. These bacteria can also use ammonia as a nitrogen source; in the presence of ammonia, they do not fix nitrogen.

The fixation of nitrogen also occurs in the photosynthetic bacteria and in some blue-green algae. These last can literally live on air and water. Nitrogen fixation has recently been found to occur in *Desulfovibrio*.

THE RELATION BETWEEN THE CELL AND ITS ENVIRONMENT

The nutrients used by a cell come, of course, from the environment; equally obviously, the nutrients are used only within the cell. An essential part of nutrition, then, is the passage of materials across the boundary of the cell.

It was shown in Chapter 3 that the cell membrane represents this boundary. We saw also that since bacteria exhibit osmotic phenomena, the membrane is not freely permeable to all substances; on the contrary, it is relatively impermeable to most.

Bacteria, in common with other cells, can accumulate substances within themselves. That is, a compound can exist within the cell at a higher concentration than without. It is then said that the cell maintains the intracellular concentration *against* the direction of the concentration gradient; simple diffusion which occurs in the *same* direction as the gradient would tend to equalize the intra- and extracellular concentrations. The accumulation of a substance requires an expenditure of energy on the part of the cell. The advantage of such accumulation to bacteria is clear. They live in dilute aqueous environments where the concentration of any nutrient may be so low as to impede metabolic reactions in which it takes part; if the cell can accumulate this nutrient, it can increase its concentration at the site of these reactions.

In many bacteria the transport of material across the membrane is facilitated by specific carriers. The membrane, in the absence of such carriers, is impermeable to a given compound. When a specific carrier for this compound is present, the membrane becomes permeable to it and the compound is rapidly transported into the cell. The outflow of the compound into the environment is apparently not facilitated by the carrier; hence, the difference in the inflow and outflow results in the accumulation of the substance. In the next chapter one such specific transport mechanism will be examined in more detail, but it should be borne in mind that the principle of specific facilitated transport is probably of very general occurrence and importance in the bacteria.

BACTERIAL ECOLOGY AND ENRICHMENT CULTURES

Ecology is the study of the interactions of organisms with their environments. Bacterial ecology has one very large handicap compared to the ecology of macroorganisms: the minute size of the environment of a microbe. The environment of any organism is of the same scale as the organism. Thus, the environment of a potted plant is the soil in the flower pot and the air around it. It is easy to analyze the soil and the air because their volumes are large. But what is the environment of a bacterium in this same soil? Clearly, a given bacterium will be in a certain cubic centimeter of soil, but that cubic centimeter of soil is *not* its environment. The environment of a bacterium—that part of the world with which it is in effective contact—is really only a few cubic micra in extent. (A cubic micron is 10^{-12} cubic centimeters.) This means that it is impossible to describe accurately the environment of a bacterium in nature. We can easily describe the properties of a cubic centimeter of soil, but we cannot describe the properties of each of the one hundred million or so microenvironments within it.

This handicap to the study of bacterial ecology is counterbalanced by the ease with which we can determine the bacteria that are best adapted to a particular environment. This determination can be made by the use of enrichment cultures. A given environment is chosen and inoculated with many kinds of bacteria (a bit of soil, for example). The form best adapted to the environment will come to the fore and can, without great difficulty, be isolated and studied. The aspect of the environment most easily manipulated is the chemical composition; one can prepare at will media of greater or lesser complexity, with various carbon sources, nitrogen sources, with or without oxygen.

A few examples will make this clear. In a well-aerated medium with hydrogen sulfide, carbon dioxide, and ammonia (in addition to inorganic salts such as magnesium, phosphate, iron, etc.), growth of one or the other of the aerobic sulfur oxidizing bacteria will occur. If the culture had been incubated anaerobically in the light, a sulfur purple or a green bacterium would have developed. If a medium containing an organic carbon compound but no nitrogen compound were incubated aerobically, *Azotobacter* would develop. A medium containing only salts and aerated with a mixture of hydrogen, oxygen, and carbon dioxide would give rise to autotrophic hydrogen bacteria.

THE ROLE OF MICROBES IN THE ECONOMY OF NATURE

The use of enrichment cultures resulted in the isolation of many kinds of bacteria with an enormous range of metabolic potentials. Some examples of these metabolisms were given in Chapter 4. Now the ways these reflect the role of bacteria as middlemen in the economy of nature must be considered.

The Cycles of the Elements

The supply of chemical elements on the earth was fixed when the earth was formed. Only a fraction of this supply is available to living things. The part of the earth that supplies organisms with the material and energy needed for their growth together with the whole mass of living matter constitute the *biosphere*. The total amount of living matter which was present over a period of time is clearly greater than it is at any one time; and, over a long period of time, exceeds the mass of the biosphere. It is clear then that the elements in the living matter must be used many times. The elements in the biosphere constantly cycle or circulate from the nonliving part into the living part by growth and back again by death and decay. Microorganisms, particularly bacteria, play essential roles in the cycles of all the biologically important elements: carbon, oxygen, nitrogen, and sulfur.

The Carbon and Oxygen Cycles

Although the supply of chemical elements on earth is fixed, there is a constant inflow of energy necessary to drive the cycles of the elements. This is the energy used in the growth of living organisms. Its ultimate source is sunlight, which becomes available to organisms through photosynthesis. Photosynthesis is also an essential step in the carbon and oxygen cycles. These two ecological functions of photosynthesis cannot be separated, since the production of O_2 and of organic compounds is the way in which the energy of sunlight is made available. Nonphotosynthetic organisms obtain energy by respiration or fermentation involving these primary products of photosynthesis. An appreciation of the magnitude of the energy used may be attained through realization that about 3×10^{18} kilocalories of light energy are converted into chemical energy each year. Since the burning of a ton of coal releases about 6,000 kilocalories, the energy trapped by photosynthesis is equivalent to the burning of about 15 million tons of coal every second.

The carbon and oxygen cycle is outlined in Fig. 5-3. The total amount of carbon fixed per year on earth is roughly 3×10^{10} tons (or about 1000 tons per second). Green plants are by far the most important agents in this process, although two groups of bacteria also contribute to it. These are the

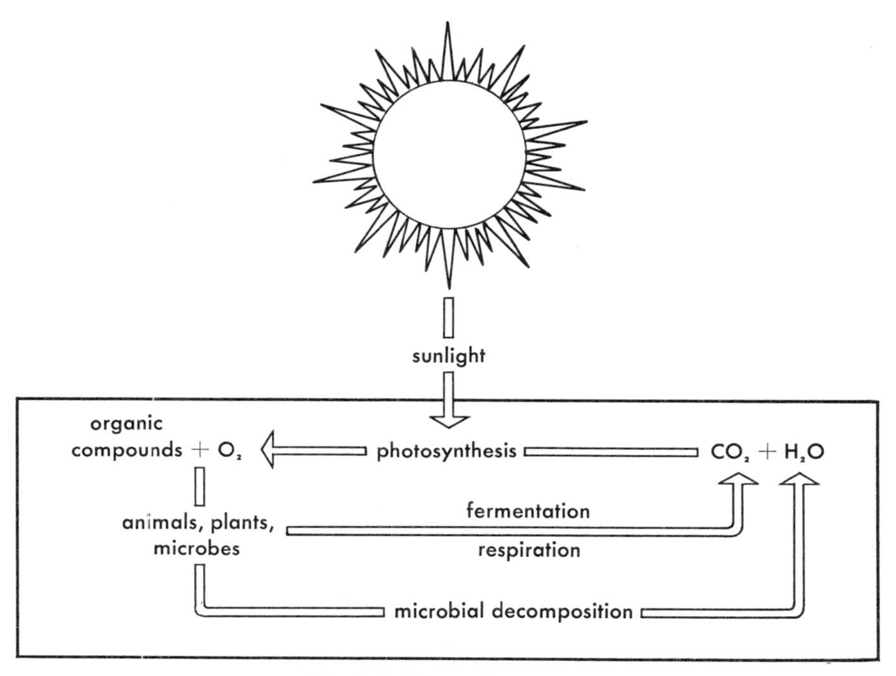

Fig. 5-3. The carbon cycle.

chemolithotrophic bacteria (see p. 46) and the photosynthetic bacteria (p. 52). It should be noted that in the case of chemolithotrophic bacteria, the fixation of carbon dioxide uses up oxygen previously formed in photosynthesis.

Bacteria play a minor role in the fixation of carbon, but they are essential to the return of carbon dioxide to the atmosphere. A large number of kinds of organic compounds are produced by plants and animals by transformations of the primary organic products of photosynthesis. All of these must eventually be converted to carbon dioxide; this is the role of bacteria. Since all naturally occurring organic compounds can be attacked by the bacteria and the fungi, all the carbon in living matter is kept in circulation. Of course, much of the carbon goes to form the bacteria themselves, but this carbon too is eventually released.

The Nitrogen Cycle

Two facts about the utilization of nitrogen by living things are important in understanding the nitrogen cycle. The first is that nitrate (NO_3^-) is a much better source of nitrogen for plants than is reduced nitrogen in the form of ammonia (NH_3^+). The second is that essentially all the nitrogen which occurs in living matter is reduced—largely in the form of amino nitrogen in proteins.

The nitrogen cycle is outlined in Fig. 5-4. Recycling of nitrogen begins with the conversion of reduced organic nitrogen to ammonia. This step is carried out by numerous microorganisms. A previously mentioned example is the fermentation of amino acids by clostridia (p. 41). The next step is the oxidation of ammonia to nitrate. How this process is carried out, in two stages, by the nitrifying bacteria has already been described (p. 47). These same bacteria also play a role in the carbon cycle. This oxidation of ammonia is the key step in the nitrogen cycle. Some nitrogen is lost from the cycle because of the action of denitrifying bacteria (p. 45), which produce nitrogen by the reduction of nitrate. Nitrogen is also lost when plants are harvested and removed from the ground on which they grew. Both these losses are made up by nitrogen fixation. A number of microorganisms are responsible for this, the most important being the symbiotic nitrogen-fixing bacteria (p. 63). In tropical areas the blue-green algae, many of which can fix nitrogen, are probably responsible for much of the nitrogen fixation.

The Sulfur Cycle

The sulfur cycle is in many ways similar to the nitrogen cycle because the most important source of sulfur is oxidized sulfur in the form of sulfate; and because, as with nitrogen, sulfur in biological material is mainly reduced. The cycle is outlined in Fig. 5-5. A large number of microorganisms can carry out the conversion of organic sulfur to inorganic sulfide. We have already

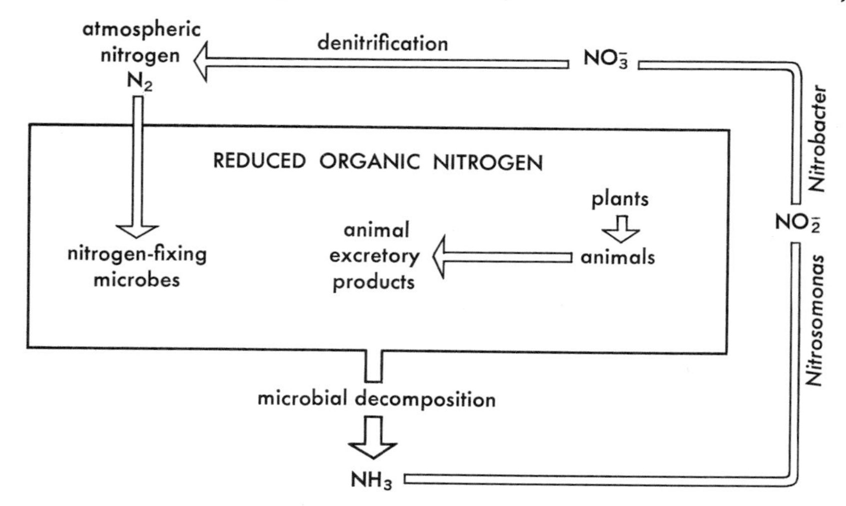

Fig. 5-4. The nitrogen cycle. The interconversions between the various forms of organic nitrogen (plant, animal, and microbial) are shown in barest outline.

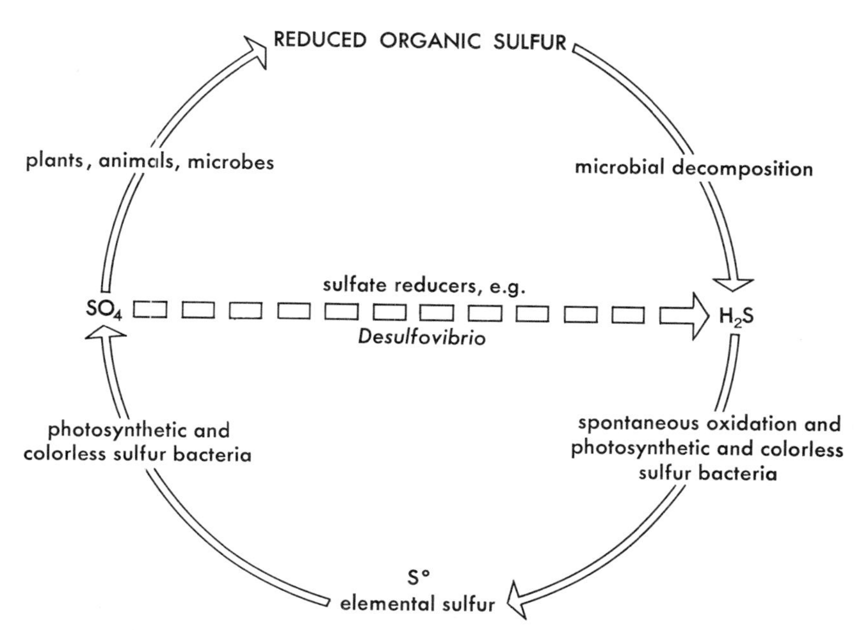

Fig. 5-5. The sulfur cycle. The bacteria that oxidize sulfide to sulfur or sulfate are discussed on pages 47 and 52, and those that reduce sulfate back to sulfide on page 45.

dealt with the metabolism of the various microorganisms involved in the other steps of the cycle.

The sulfur cycle is probably not so critical as the nitrogen cycle, since there is an abundant supply of sulfate in the earth's crust. The interconversions of the different forms of sulfur in the cycle are of considerable geochemical importance. For example, many of the large deposits of elemental sulfur were formed by the action of microorganisms.

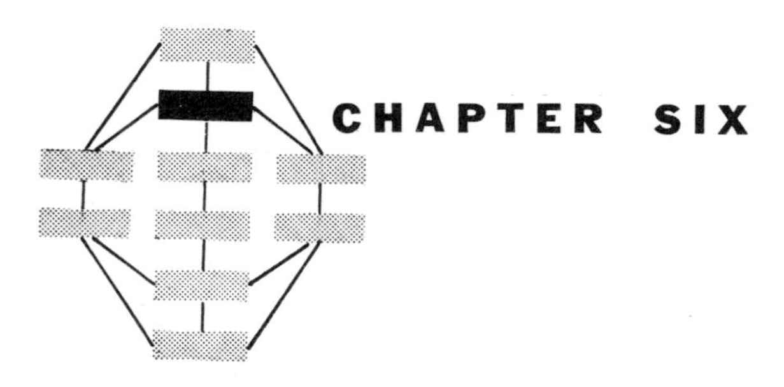

CHAPTER SIX

GROWTH AND PROTEIN SYNTHESIS

In the two previous chapters the problem of how a bacterium makes itself was approached by a study of the reactions necessary for energy metabolism and for the biosynthesis of small molecules which underlie the growth of the cell. It was shown how the cell captures chemical or light energy and how this energy is used in biosynthetic reactions.

How does the cell use these small molecules to make the nucleic acids, proteins, and polysaccharides which characterize it? As yet little is known about the individual reactions of macromolecular synthesis or how these reactions work together to produce a new cell. However, a good deal can be learned by analyzing the resulting growth and protein synthesis. This chapter will take up first the growth of populations of bacteria and of the individual cell, and then the formation of enzymes by bacteria. Some of the information in this chapter has been covered in *Cell Structure and Function* in this series; it would be worthwhile to review this material before going on.

THE GROWTH OF POPULATIONS OF BACTERIA

The Mathematics of Growth

To begin with, let us define *growth* simply as an increase in mass of bacteria per unit volume of medium, usually accompanied by an increase in the number of cells. All the constituents of the bacterial mass must come from the medium. Material flows into the bacteria from the medium, part of it is excreted back into the medium as the end products of the energy metabolism of the cell, and the rest is converted by biosynthetic reactions into new cell material. Part of the bacterial mass is composed of the intermediates in these reactions—the metabolites of the cell. The larger part comprises the macromolecules, which are the end products of all these reactions. A culture of bacteria cannot be thought of simply as a factory into which raw materials enter and from which products emerge; the essential point of growth is that the product is more factory. In other words, the whole complex pattern of enzymes, metabolites, nucleic acids, lipids, polysaccharides, and so forth is increasing as a whole. Not only are all these things increasing, they are all increasing at the same rate. If they were increasing at different rates then the cell would not only be growing it would be changing. All this can be said more simply: bacteria are self-replicating.

The increase in mass of any self-replicating system is autocatalytic. In other words, the rate of increase of new bacteria is proportional to the mass of bacteria. (It may be pointed out that all autocatalytic processes are not self-replicating; an explosive chain reaction is a good example.) This concept can be stated in terms which are perhaps somewhat more familiar. The birth rate of a human population, that is, the rate of formation of new individuals, is proportional to the number of individuals in the population. The specific growth rate is simply the birth rate per individual. For example, if in a city of one million inhabitants the birth rate is ten thousand per year, then the specific birth rate is 10,000 per year per 1,000,000 or 1/10,000 per year. In a similar way we can define the specific growth rate of a bacterial population. If the total rate of synthesis of new cell material at any time is, say, 100 μg per ml per min[1] and if the total mass at that time is 1000 μg per ml, then the specific growth rate is 100 μg per ml per min divided by 1000 μg per ml or 0.1 per min. We could, of course, have expressed growth in terms of cell numbers; the growth rate would have been the same. We will use the symbol α to denote the specific growth rate. Note (and ponder) that α is given in units of reciprocal time, for example, per hour, or per minute.

It is characteristic of the growth of self-replicating units that in a constant environment and after an initial period of adjustment, the specific growth rate tends to become constant. During the period of adjustment, the specific growth rate is increasing. When the specific growth rate is constant, the rate at which new cell material is being formed at any time is simply the mass of cell material present times the specific growth rate. This follows obviously from the definition of specific growth rate; α = total rate of increase per unit mass of cell material. In other words, the population will increase by a constant factor over equal time intervals. This factor is simply the specific growth rate times the length of time interval (care being taken to see that the same units of time are employed). If initially the mass of bacteria is B_0 then after one time interval, the mass will be $B_0 \times \alpha t$ or B_1. After a second equal interval, it will be $B_1 \times \alpha t$ or B_2; in terms of the initial mass, $B_2 = B_0 \times \alpha t \times \alpha t$ or $B_0 \times (\alpha t)^2$. Similarly, after the third interval

[1] 100 μg (micrograms) is one ten-thousandths of a gram.

$B_3 = B_0 \times (\alpha t)^3$, and so on. In considering the growth of a self-replicating system it is reasonable to take as the unit time interval the time required for the mass to double, or in other words the time needed for the system to replicate itself once. This time is called the *mass doubling time*. If we use the doubling time as our time unit, the factor increase, αt, per unit time is clearly 2. Therefore, the simple equations developed above become: $B_1 = B_0 \times 2$, $B_2 = B_0 \times 2^2$, and $B_3 = B_0 \times 2^3$. Now the power to which the number 2 must be raised to equal any number is called the logarithm to the base 2 of that number. (If $X = 2^Y$, then $\log_2 X = Y$.) Our equations can be put in logarithmic form as follows: $\log_2 B_1 = 1 + \log_2 B_0$, $\log_2 B_2 = 2 + \log_2 B_0$, and $\log_2 B_3 = 3 + \log_2 B_0$. After any number, N, of doublings, we can calculate the mass of bacteria by the formula $\log_2 B_N = N + \log_2 B_0$. It should be clear that if the logarithms to base 2 of the mass of bacteria is plotted against time (still expressed in doubling time units), a straight line will result, as shown in Fig. 6-1. A glance at this graph will make the following points clear. It is obvious that the slope of the line is equal to the increase per unit time of the logarithm of the bacterial mass. If time is in doubling time units and logarithms to the base 2 are used, the slope will be unity, since the mass doubles in one unit of time. The slope of the line is just another way of expressing the specific growth rate constant, α, in terms of doublings per unit time. If minutes are used as the time unit, then the slope of this line is clearly 0.01 and α is 0.01 doublings per minute.

The arithmetic is a little less obvious if logarithms to the base 10 are used. Reference to the figure shows that in this case a doubling of the mass increases the logarithm by only 0.3, that is, $\log_{10} 2$, and this is the slope of the line with time expressed in units of doubling time. The slope is obviously 0.003 when the time is given in minutes. These slopes give the value of α in the equations when common logarithms are used.

$$\log_{10} B_N = \alpha t + \log_{10} B_0$$

It is easy to get the value of α in doublings per unit time by dividing by $\log_{10} 2$, which is 0.3.

In general, if the specific growth rate is constant, the logarithm (to any base) of the mass of bacteria will increase linearly with time. The slope of this line is equal to α, and the numerical value of α depends upon the time unit used and the base of the logarithm.

For those of you with a little mathematical training all of this can be put more succinctly. For the rate of increase of bacterial mass we can write:

$$\frac{dB}{dt} = \alpha B$$

If α is constant, the equation can be integrated to give:

$$Bt = B_0 e^{\alpha t}$$

or in logarithmic form:

$$\ln B_t = \ln B_0 + \alpha t$$

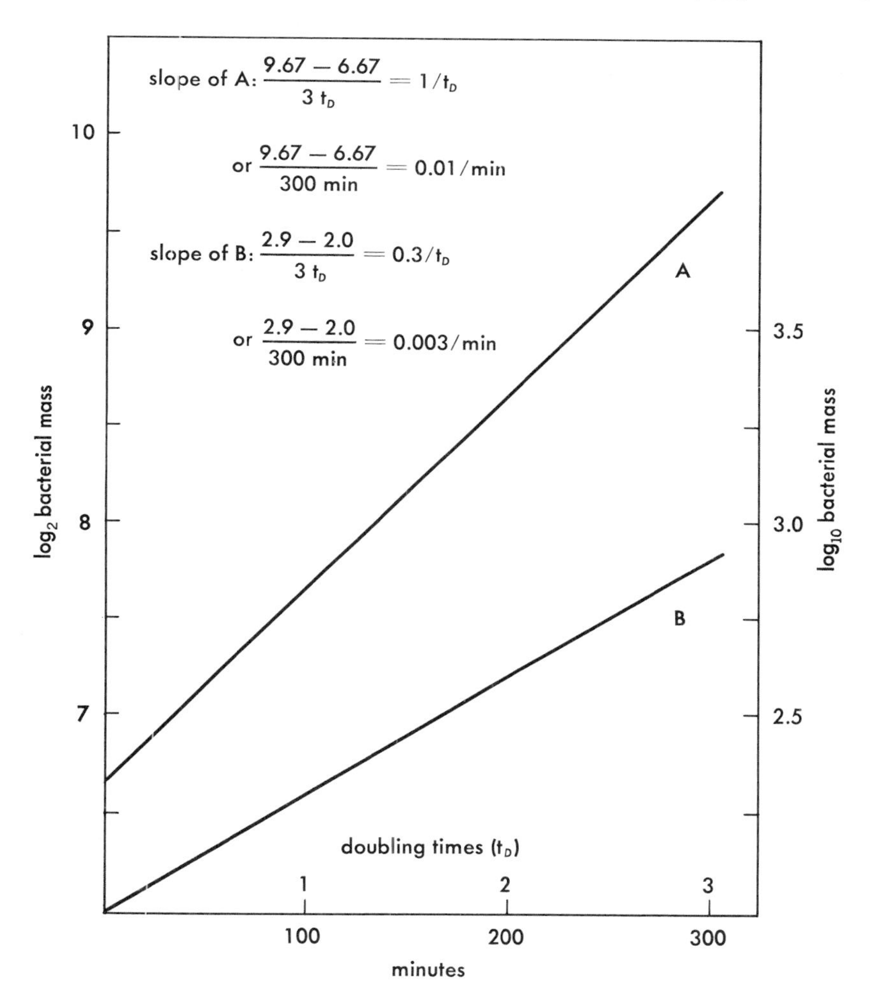

Fig. 6-1. Exponential growth. Curve *A* shows the increase in the logarithm to the base 2 of the bacterial mass, and curve *B* the increase in the logarithm to base 10. The slope of curve *A*, when the time for one doubling is taken as the time unit, is unity. In this example, the doubling time is 100 minutes, hence the slope of curve *A* is 0.01 per minute.

The numerical value of α when natural logarithms are used, as here, is again different from that in the equations using logarithms to the base 2 or 10. It can be seen from these last equations that the most natural time unit is one in which αt will be equal to 1 when t is 1; in other words, a unit of time of $1/\alpha$. When this unit is used, during every unit increase in time the mass increases by a factor equal to base of logarithms used: e, 2, or 10. This then, is what we did originally when we took the doubling time as our time unit and used logarithms to the base 2.

It was said that the specific growth rate can remain constant only in an unchanging environment; a little thought will show that this condition can never be fulfilled in nature. The environment is being changed, by growth, into bacteria; any limited environment must change, and change more rapidly as more growth occurs. It can be expected, therefore, that the specific growth rate will eventually begin to decline, and that finally growth will cease, either because toxic products have accumulated or because the medium has become exhausted.

The Growth Curve

With this mathematical background we are now in a position to understand what goes on during the growth of a bacterial population. Let us see what happens when a certain amount of medium is inoculated with a known amount of bacteria and incubated under constant physical conditions. The results of such an experiment are shown in Fig. 6-2. Let us consider the various parts of this figure in detail.

At first, there is a period during which little if any growth occurs. This is called the *lag phase.* Sooner or later the bacteria begin to increase, at first slowly but with an increasing specific growth rate which eventually becomes constant. The lag phase and the period of increasing specific growth rate are a time of adjustment of the cells to the new environment. During the lag phase the cell mass usually increases somewhat while the number of cells remains constant; in other words, the cells become larger. The length of this phase is extremely variable. What may be going on inside the cells during this time will be discussed subsequently.

The next phase is characterized by a constant specific growth rate. This is called the phase of exponential or logarithmic growth. The value of the specific growth constant depends upon both the bacterium and the medium; it is a constant only for a particular bacterium growing under a particular set of conditions. Between the lag phase and the exponential phase is a period (usually rather short) of increasing growth rate. This is shown more clearly in the small figure.

During exponential growth in a constant environment, the physiological properties of the bacteria are constant. One of the most important of these constant properties is the ratio of mass to area. Since this ratio is constant, it follows that cell division keeps pace with increase in mass. The cells become larger during the lag phase, reach their maximum size at the end of the lag phase, and remain the same size all during the exponential phase.

It is important to understand that cell division is not the *cause* of exponential growth. However, it is necessary for continued exponential growth, since it is the way in which bacteria maintain a constant ratio of mass to area. Equally important for continued exponential growth is the

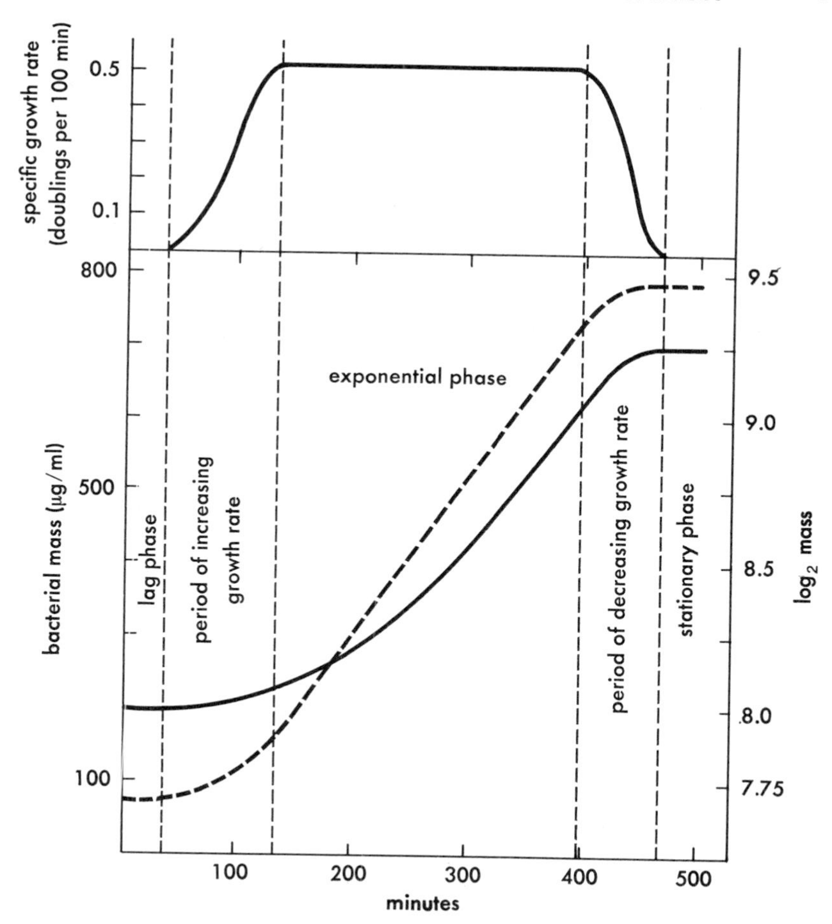

Fig. 6-2. The bacterial growth curve. The lower part of the figure shows how the bacterial mass and the logarithm of the mass increase with time. Note that the logarithms are to base 2. The upper part of the figure shows how the specific growth rate changes; it is constant only during the exponential phase of growth.

maintenance of a constant cell composition. This means that the relative amounts of enzymes, metabolites, and other constituents do not change during exponential growth. The term balanced growth is used to denote this constancy of physiological characteristics. If the environment changes, the physiological state of the cells will change. Usually this change will be simply a transition to a different constant physiological state. If the change in the environment is such that the cell cannot achieve a new state of balance, growth becomes permanently imbalanced. If this condition continues long enough the cells die. A striking example is provided by the action of penicillin. This antibiotic specifically inhibits the synthesis of the bacterial cell wall. The rest of the cellular components continue to be synthesized, but not exponentially, inasmuch as all parts of the cell are not being replicated. Since formation of a new cell wall is essential for cell division, fission does not occur and the ratio of mass to area cannot be kept constant. Eventually, the cell wall can no longer retain the mass of growing cell material within it. The wall then ruptures and the cell lyses.

This extreme example of unbalanced growth provides a clue to the mechanism by which the specific growth rate does not increase without limit. All parts of the network of reactions underlying growth are interrelated. Therefore, no one part of the network, for example that involving the synthesis of some one particular building block, can be "outgrown" by the rest, for the rate at which this one building block is supplied will limit the rate at which the rest of the network functions. If, in this example, the building block were supplied from the medium, then that particular limitation would be circumvented and the specific growth rate would increase until some other part of the network became limiting. Note that the physiological state of the bacteria will be different in the two cases; in particular, the concentration of the limiting building block will be higher in the cells with the greater specific growth rate.

As already stated, it is impossible for exponential growth to continue indefinitely in a finite volume of medium; the environment must change and with it the growth constant. As shown in Fig 6-2, growth eventually stops; there is no further increase in mass. The culture has entered the *stationary phase* of the growth curve. The stationary phase is preceded by a period of declining specific growth rate. In the stationary phase the proportionality between bacterial mass and cell number may again break down. Some bacteria simply stop growing, with the number of viable cells remaining more or less constant for rather a long time—much longer than the doubling time of the growing cells. The size of the cells, on the other hand, decreases somewhat. Certain bacteria begin to die when growth ceases.

The stationary phase is not the result of any kind of intrinsic aging of the bacteria. This fact can be easily shown by transferring a portion of a culture in the exponential phase of growth to fresh medium; the diluted culture continues to grow exponentially. This can be repeated indefinitely. Only the medium ages, not the cells. It is meaningless to speak of young or old cells; only the culture (cells plus medium) can be spoken of as old or young.

The medium can change in two ways. In the first place, metabolism invariably results in the excretion of waste products and as these accumulate the medium may become toxic, thereby reducing the growth rate; eventually growth ceases. The second way in which the environment can change and limit the amount of growth is for the medium to become exhausted of some essential nutrient. For example, since nitrogen makes up about 10 percent of the dry weight of bacteria, a medium that contains 100 μg of nitrogen per ml can yield no more than 1000 μg of bacteria per ml. If the nitrogen content of the medium were only 50 μg per ml, only 500 μg per ml of bacteria could be produced. A similar situation obtains with a nutrient that serves as both carbon and energy source. Here, however, more of the nutrient is used for growth than can be accounted for as cell carbon; the rest is excreted as end products of energy metabolism. It is clear that in the case of a nutrient such as nitrogen, the relationship between the amount of nutrient supplied and the amount of cells formed is that shown in Fig. 6-3. This curve, of course, passes through the origin, since without nitrogen there is no growth. The same kind of relationship is found between growth and the amount of energy source provided. The curve in Fig. 6-3B, which relates the amount of growth to the amount of glucose provided, also passes through the origin. This means that growing bacteria use all of the available energy for growth. The only thing they do is grow.

The concentration of a limiting nutrient can influence not only the amount of growth but also the growth rate, as shown in Fig. 6-4. The growth rate is maximal and independent of the concentration over a wide range of concentrations. It starts to decline only when the concentration of limiting nutrient is very small. Usually the initial concentration of the nutrient will be very much greater than that at which the growth rate begins to decline. The growth rate does not decrease until nearly the maximum amount of growth is attained. Thus, the period of declining growth rate is very short.

THE CELLULAR GROWTH CYCLE

So far the discussion has concentrated on the growth of populations of bacteria rather than on the growth and division of individual cells. By itself, knowledge about the growth of populations of bacteria is of limited interest; its value is in what it can tell us about the growth of the individual cell. Unfortunately, the growth of a single bacterium cannot be studied in detail for the simple reason that it is too small. We can find out how long it takes for a single cell to divide by observing it under the microscope, but this is about as far as direct observation of cells can go. The microbiologist then is forced to use populations to find out about individuals. Clearly, there is some relationship between the way in which individual cells grow and the way in which a population of many similar cells grows, but this relationship is not necessarily a simple one. For example, the exponential growth of a population does not imply that the growth of a single cell is exponential. Only some questions about the growth of individual cells can be answered by studying populations; indeed, in some cases, populations give false answers to questions about cellular growth.

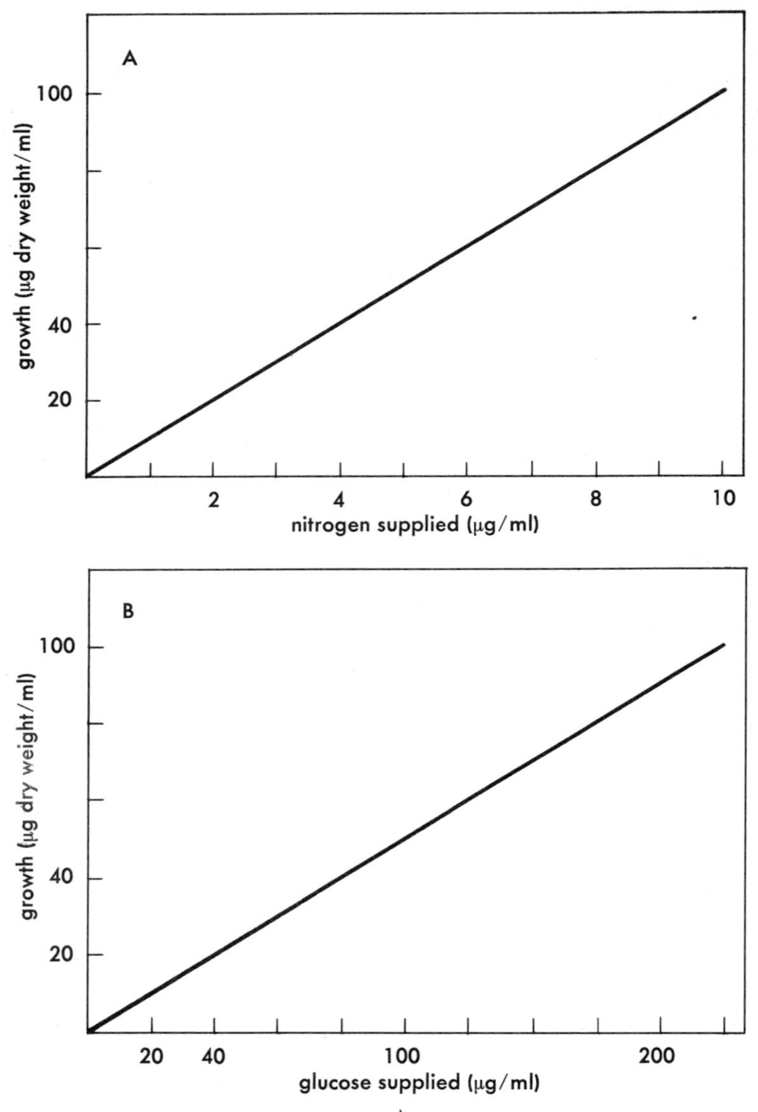

Fig. 6-3. Growth yields. A: How the amount of bacterial mass increases as the supply of the limiting nutrient increases. In this case the limiting nutrient was nitrogen. The slope of the line (100 μg of bacteria per 10 μg of nitrogen) shows that bacteria are 10 percent nitrogen. B: How the growth yield increases as the supply of the carbon and energy source increases. Notice that the line passes through the origin, indicating that all the energy from the metabolism of glucose is used for growth. If energy were needed for other purposes the curve would have shown a small amount of glucose needed for no growth.

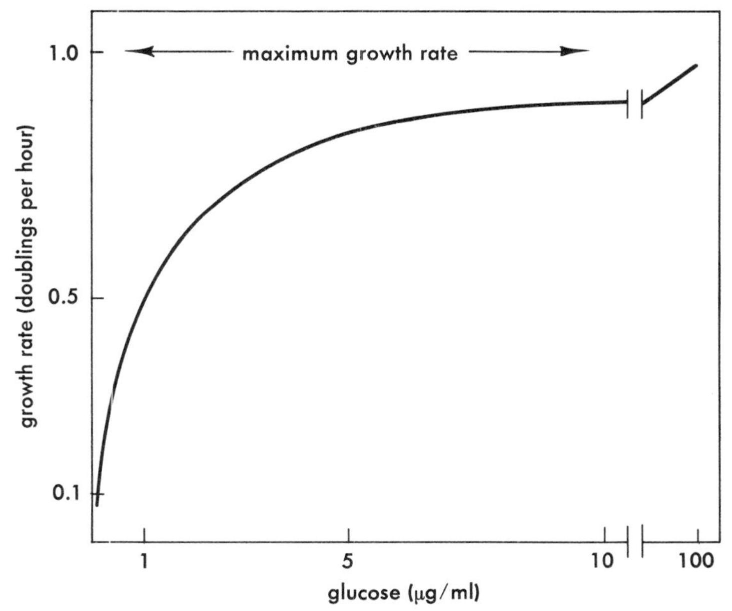

Fig. 6-4. The influence of the concentration of a limiting nutrient on the specific growth rate constant. The growth rate constant depends upon the concentration only at very low concentrations. A comparison of this figure with Fig. 6-3B shows that if initially 1000 μg of glucose was provided, the specific growth rate would remain at its maximum value until more than 90 percent of the total growth had occurred.

Heterogeneity of Populations of Bacteria

The simplest question which can be asked about the growth of a bacterium is "How long does it take to double its size and divide?" As was just mentioned above, this question is easily answered by observing growing cells under the microscope. It is found that the cells in a population do not have identical division times, but that division times are distributed around an average in the familiar bell-shaped normal distribution curve. This is illustrated in Fig. 6-5, which shows the distribution of division times for two hypothetical populations.

The value of the *mean* division time is the same for both populations, but the "spread" of individual division times is very different in the two. The two populations would have the same mass doubling time (30 minutes) if this were calculated from a growth curve according to the equation on page 72. By studying the population as a whole we can learn what the mean division time is, but we cannot learn anything about the individual generation times or how these are distributed. This is the first example of the kind of information not obtainable from a study of populations.

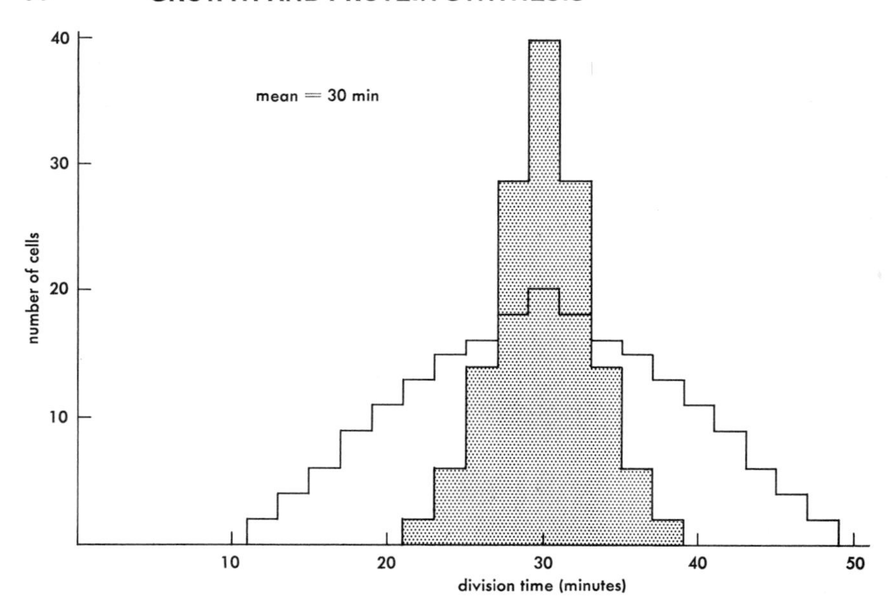

Fig. 6-5. The distribution of division times, showing the frequency with which different division times occur in populations of two different bacteria. The mean division time is the same for both, but the distributions are very different.

The heterogeneity of division times explains why a population descended from a single cell does not grow in a stepwise fashion. In the development of a population from a single cell, the first few divisions are synchronous or nearly so; that is, all the cells divide at about the same time. But after a few generations, the divisions will begin to be out of phase with one another. By the time the population is large enough to be studied easily, division is completely random. Thus, at any one time every stage in the cellular division cycle is represented in the population.

As stated previously, during the exponential phase the growth of a population is balanced, or in other words, the chemical composition and physiological properties of the population are constant. However, since growth is random, it cannot be inferred that the growth of a single cell is balanced. Indeed, it is known that in large cells such as amoebae, the chemical composition of the cell changes during the division cycle. Balanced growth means that at corresponding points in the division cycle the properties of a cell and its progeny are the same. The constant properties of a randomly growing population are the same as the properties of a single cell averaged over its division cycle. Such averages are clearly of little value in understanding the growth of the individual cell.

Recently it has become possible to obtain large populations of bacteria that are growing synchronously; that is, at any one time all the cells are in the same stage of the division cycle. It seems certain that studies of such populations will go far toward answering many questions about the cellular growth cycle of bacteria.

The mean division time is one property of the single cells that can be determined from a study of populations, because the distribution of individual division times is symmetrical. In other words, the number of cells with division times shorter than the average is the same as the number of cells with longer than average times. Since all properties may not be symmetrically distributed like this, however, an analysis of the population may simply give false answers. As an example, consider what happens during the formation of certain induced enzymes (below). Initially, the culture has no enzyme at all. Then, after an appropriate stimulus, the enzyme begins to be formed. Let us say that after 30 minutes the culture has E units of enzyme. If there are N cells in the culture, it might be assumed that each cell has, on the average, E/N units of enzyme. This assumption is reasonable but completely false; in fact, some cells have no enzyme at all and some have much more than the average. In the population shown in Fig. 6-6, about one half of the cells have from nine to ten units of enzyme per cell, and the rest have zero to three units. The average (total enzyme units divided by the number of cells) is about 5.3 units per cell, but actually no individual cell has this amount. The statement that the population shown in Fig. 6-5 has an average division time of 30 minutes clearly is more meaningful than the statement that the average enzyme concentration is 5.3 units per cell in the population of Fig. 6-6.

It is often said that one of the useful properties of bacteria as a biological tool is that it is easy to obtain large numbers of them and so it is possible to get a statistically good measurement by a single determination. The example just quoted shows that a naïve acceptance of this concept can lead one seriously astray.

THE SYNTHESIS OF PROTEINS BY BACTERIA

So far in this book the growth of bacteria has been handled in a purely quantitative way—how fast cell material is being formed. The question of how the cell regulates the *kind* of material it makes has not yet been considered. We must turn now to this question, which is one of the central problems of modern biology.

Induced Enzyme Formation

You are already aware from your study of genetics that what a cell looks like (its phenotype) is determined ultimately by the genetic make-up of the

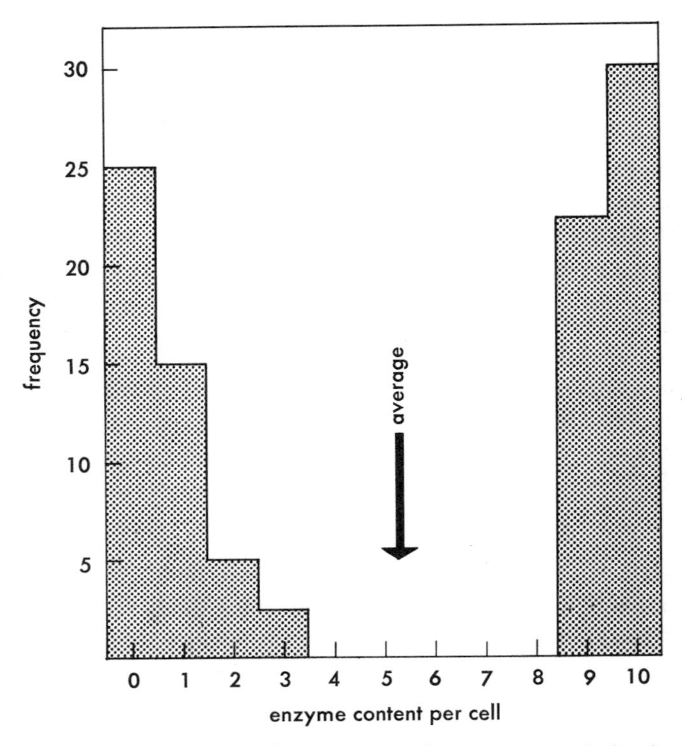

Fig. 6-6. Discontinuous heterogeneity, showing in an idealized way, how the amount of an enzyme per cell can be distributed in a population. Notice that the average amount calculated from the total amount of enzyme and the total number of cells does not correspond to any actual amount.

cell (its genotype). However, the phenotype is not determined solely by the genotype, but rather by the interaction of the genotype and the environment. In complex organisms the "environment" is, to a large extent, the product of the genotype; the environment of the kidney cells, for example, is clearly not the same as the environment of the human body as a whole. Microbes, especially bacteria, are in much closer contact with the environment; the phenotype of a bacterium is, accordingly, much more plastic than that of other organisms. In other words, in a bacterium the importance of the environment in determining phenotype is relatively greater than in more organized forms of life. As might well be expected, most of the phenotypic plasticity of bacteria is due to their abilities to synthesize different proteins, especially enzymes, in response to various stimuli. We will consider in the next few pages just how and under what conditions a bacterium synthesizes new kinds of enzymes. This type of study has proved to be of enormous significance to biology as a whole since through it much has been learned about the way in which proteins are synthesized. These ideas are discussed at length in

Cell Structure and Function in this series, hence the emphasis here will be more on the significance of these mechanisms to the biology of the bacteria.

The following example will be considered in some detail. *Escherichia coli* is a bacterium that lives in the intestinal tract of higher animals. It is a typical organotroph; it will grow on a medium containing a single organic compound as energy and carbon source plus the usual inorganic salts. *E. coli* can attack a wide variety of organic compounds, including the sugars glucose and lactose. It is the physiology of the metabolism of glucose and lactose with which we will be concerned.

Occasionally a type (or strain, as microbiologists would say) of *E. coli* that cannot grow on lactose is isolated. This strain may be referred to as lactose (−); the more usual type is lactose (+). The difference here is genetic; lactose (−) strains are mutants that cannot form all the enzymes necessary for lactose metabolism. A lactose (−) strain can also mutate to yield a lactose (+) strain.

Lactose is a disaccharide consisting of a molecule of glucose linked to a molecule of galactose.

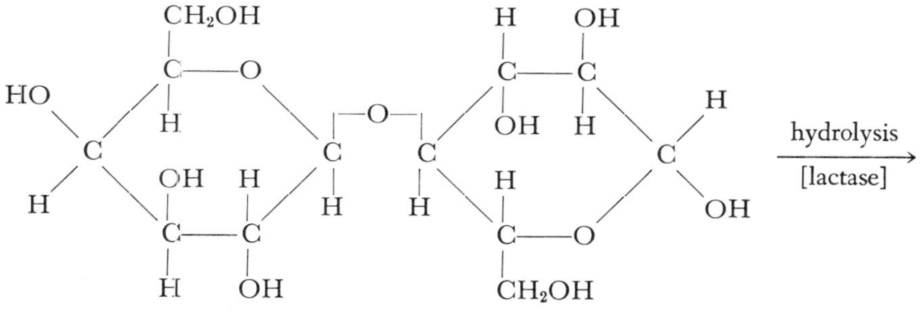

CH_2OH H OH
C—O C—C
HO H H OH H
C C —O— C C hydrolysis
H OH H H H H OH [lactase]
C—C C—O
H OH CH_2OH

lactose: galactoside-glucose

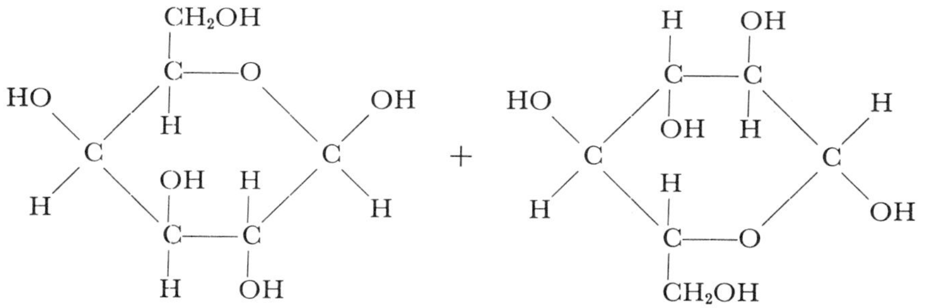

CH_2OH H OH
C—O C—C
HO H OH HO OH H H
C C + C C
H OH H H H H OH
C—C C—O
H OH CH_2OH

galactose glucose

All strains of *E. coli* can grow on glucose; therefore, the only *necessary* difference between lactose (+) and lactose (−) strains is the ability of the former to obtain glucose from lactose. Lactose (+) strains possess an enzyme, called lactase, which catalyzes the splitting or hydrolysis of lactose into glucose and galactose. In addition to lactase, a lactose (+) strain must be permeable to lactose. This is so because *lactase* is not excreted into the medium; a strain that forms lactase but is not permeable to lactose is not able to grow on lactose. We can symbolize the ability to form lactase by Z^+ and the inability by Z^- and, similarly, for the permeation system, Y^+ and Y^-. There are then four possible combinations:

Z^+Y^+ the normal (wild) or lactose (+) type; forms lactase, is permeable to lactose.

Z^+Y^- does not grow on lactose; forms lactase, but is not permeable to lactose.

Z^-Y^+ does not grow on lactose; does not form lactase, but is permeable to lactose.

Z^-Y^- does not grow on lactose; does not form lactase, and is not permeable to lactose.

Each of these four possible types of *E. coli* has been isolated.

These differences are all genetic and a lactose (−) strain can become lactose (+) only by mutation. However, it is found that lactose (+) or Z^+Y^+ strains contain lactase and are permeable to lactose only *after they have been grown on lactose.* When growth has been at the expense of glucose (or any other carbon source) the cells do not contain any lactase. If these cells are now placed in a medium in which lactose is the sole carbon source, they will have a rather long lag period before growth begins. During this lag period, the enzyme lactase (as well as the specific permeation system) are being synthesized. The "lag phase" in the growth curve (discussed earlier) may often be caused by such a response.

An enzyme which shows this effect is called an adaptive or induced enzyme. Many of the compounds used by bacteria are metabolized by induced enzymes. A few examples will illustrate the generality of the phenomenon.

Many bacteria can grow using aromatic compounds, such as benzoic acid or naphthalene, as the sole source of carbon and energy. The metabolism of a single one of these compounds may involve three or four or more enzymes, each one of which is induced.

Denitrifying bacteria (see p. 45) form the enzymes necessary for the reduction of nitrate only when nitrate is present.

Before the significance of induced enzymes to the biology of the bacteria is considered, the way in which *E. coli* forms lactase should be examined in some detail.

In the first place, the ability of the cells to split lactose is the result of an increase in the amount of lactase protein in the cells. The increase in the activity does not result from the *activation* of a previously present but enzymically inactive protein, but rather the cells, after the addition of lactose, begin to make a new protein not previously made.

In *Cell Structure and Function* it is pointed out that enzymes are very specific in their action. It might be thought that an enzyme could only be induced by its specific substrate. This, however, is not true; a variety of galactosides other than lactose can elicit the formation of lactase. Lactose and these other galactosides are all called inducers of lactase. Regardless of the inducer, the enzyme formed is the same. Some of the inducers are not attacked at all by the enzyme which they induce. These nonmetabolizable inducers have been a tremendous help to the study of induced enzyme synthesis. When such inducers are used, a carbon and energy source other than lactose must, of course, be provided. Under these conditions it is found that immediately after addition of the inducer, the cells begin to make lactase. From the moment of induction the enzyme is a constant fraction of the new cell protein formed. In other words, the rate of enzyme synthesis with respect to total protein synthesis is constant. However, since the cells are growing exponentially the rate of enzyme formation with respect to time is not constant. To make this very important idea more concrete, let us assume that we have a culture of *E. coli* containing 100 μg of cell protein per ml; let us assume, further, that after addition of the inducer, 5 percent of the *newly* formed protein is lactase. Thus after one generation there will be a total of 200 μg of protein, of which 100μg will have been made since the addition of inducer; hence, there will be a total of 5 μg of lactase, which is 2.5 percent of the total protein (old and new). After two generations there will be 400 μg of total protein, of which 300 μg are new; hence, there will be 15 μg of lactase, representing 3.75 percent of the total protein. After four or five generations the protein which was present at the start will be but a small fraction of the total, and the lactase will approach 5 percent of the total protein.

A culture behaves in this fashion only when the inducer is present in a relatively high concentration; the result is different at very low inducer concentrations. Here, the rate of enzyme synthesis relative to total protein synthesis is not constant. On the contrary, the rate at first is low and increases to the same rate that is attained immediately with high inducer concentrations. This is because the permeation system necessary for the rapid entry of substrates and inducers of lactase must itself be induced. In uninduced cells, which lack the permeation system, the inducer molecules can enter only by diffusion. When the inducer is present in low concentration, the probability that a few molecules will enter a cell is small. But once the entry of molecules begins, the specific permeation system is induced and the rapid entry of more inducer is allowed. Thus, a cell goes from a state of non-induction to one of complete induction in a short time. The increasing rate of enzyme synthesis relative to total protein synthesis reflects an increase in the *number* of cells that are induced and not the *extent* to which the whole population is induced. This is the kind of heterogeneity discussed on page 81.

The Irreversibility of Bacterial Growth

So far, the discussion has been on the formation of specific proteins in bacteria; we have assumed that a culture of *E. coli* could be obtained *without* any lactase. But how is this done? In other words, what happens to the lactase when the lactose is removed and replaced with another carbon source? The answer is simple and profound: nothing happens. The cells, of course, continue to grow on the new carbon source and lactase stops being formed, but it is not broken down. The amount of enzyme in a given *volume* of culture stays constant, but the amount of enzyme per cell decreases. Continued growth simply dilutes the enzyme. It follows that in a growing bacterial cell the proteins, once formed, are not broken down. By sharp contrast, in cells of higher organisms the protein is constantly being degraded and resynthesized. In other words, in bacteria none of the energy expended in protein synthesis is wasted.

Under certain circumstances bacterial proteins are not stable; for example, when cells are deprived of a carbon source, they begin to break down their proteins. If cells devoid of lactase are put in a medium in which lactose is the only carbon source, they can initiate the formation of lactase in the absence of any increase in the total amount of protein. Where do the amino acids required for lactase synthesis come from? The amino acids are produced by slow breakdown of the cell proteins. Most of the amino acids freed in this way are used to remake the same kinds of proteins. However, a small fraction of the amino acids are used to synthesize the lactase, a *new* protein, if an inducer is present. The proteins in nongrowing cells are said to be "turning over." The advantage of this property to the bacteria is obvious: without this ability to form new kinds of protein, the cells would be unable to grow in the new medium.

The Control of Protein Synthesis

It has been shown that the ability to form an induced enzyme is under strict genetic control. A cell that is genetically incompetent to form the enzyme will not form it under any circumstances; a cell which is competent forms the enzyme only under certain circumstances.

We can think of two ways in which the synthesis of an enzyme can be influenced by the genetic make-up of a bacterium. The *structure* of the enzyme protein is genetically determined: the sequence of the amino acids and the way in which the protein molecule is folded are controlled by a particular portion of the genetic material. Mutations here lead to structurally modified proteins; in most cases these changed proteins are enzymically inactive. The formation of such altered enzymes can be shown only by sensitive immunological techniques. The second way in which the synthesis of a protein can be modified genetically is by a change in the conditions under which the cell makes enzymes. The first evidence for the existence of this kind of genetic control was the discovery of mutants of *E. coli* that form lactase in the absence of an inducer. Such mutant strains are said to be *constitutive* for lactase, in contrast to the *inducible* strains in which lactase is formed only in the presence of an inducer. Many enzymes, most notably those concerned in biosynthetic reactions, are normally constitutive. We have just seen that small molecules can induce the synthesis of inducible enzymes. A compound that is an inducer of lactase in inducible strains has no effect upon its synthesis in constitutive strains. However, one of the *products* of the enzyme reaction, galactose, inhibits the synthesis of lactase in constitutive strains. This inhibition of synthesis is even more striking in the case of biosynthetic enzymes. For example, cells grown in the presence of the amino acid tryptophan do not form the enzymes involved in the biosynthesis of tryptophan. Just as it is possible to obtain mutants that form lactase in the absence of inducer, so is it possible to obtain mutants that, for example, synthesize the enzymes of tryptophan biosynthesis in the presence of this amino acid.

Clearly, the regulation of protein synthesis is under genetic control, as is the structure of the protein. The structure of a protein is determined by a *structural gene,* the regulation of its synthesis by a *regulatory gene.* Each regulatory gene influences the synthesis of a specific enzyme. It is thought that the regulatory gene produces a substance that can inhibit the action of the structural gene. For example, in a cell that is inducible for lactase, the product of the regulatory gene is always present and synthesis of lactase is inhibited. The inducer acts by relieving this inhibition in some way; how this happens is not yet clear. In the regulation of other enzymes, such as those concerned in tryptophan biosynthesis, the substances produced by the regulatory genes can act only in the presence of certain compounds, such as tryptophan. If a mutant occurs in which the regulatory gene has been affected, then enzyme synthesis occurs under all conditions. These are constitutive mutants.

Recently it has been found that not all constitutive mutants are of this sort; in some, the structural gene is able to act in the presence of the inhibitor formed by the regulatory gene. In other words, the structural gene has become insensitive to the inhibitor.

This method of controlling the protein-forming machinery of the cell may seem unnecessarily complex. Before some of its physiological advantages to a cell are demonstrated, the machinery itself will be briefly described.

Cell Structure and Function discusses the importance of ribonucleic acid to protein synthesis in other kinds of cells. This compound is equally important in the bacteria. The ribonucleoprotein particles, or ribosomes, discussed in Chapter 3, seem to be the sites of protein synthesis. It should now be clear why we separated the protein associated with these particles from the rest of the cell protein. A ribosome, by itself, is not able to synthesize a specific protein. To do so it must be "activated" by that portion of the deoxyribonucleic acid (DNA) that determines the structure of the protein. The way in which this activation occurs is just now beginning to be understood. The important role of ribosomes in the synthesis of proteins is very clearly shown by the fact that the number of ribosomes per cell depends upon the rate at which the cell is growing. The growth rate of a bacterium, such as *E. coli,* that can make all its own building blocks can be greatly increased by providing the cells with these compounds in the medium. Now, a cell doubling every 20 minutes is clearly making protein at a higher rate than one doubling every 80 minutes, and the number of ribosomes in the faster-growing cells is greater than that in the slower-growing cells. To make protein faster, a cell must increase the number of its protein factories—or ribosomes.

Since the number of factories can limit the growth rate, it is clearly an advantage to the cell to use them efficiently and economically. The kind of control discussed earlier insures the efficient use of the protein-forming machinery of a bacterial cell. Let us see how this control might operate in a growing cell. Recall here the network of reactions by which the cell synthesizes its building blocks and how this network is connected by a number of common intermediates with the reactions of energy metabolism. Fig. 6-7 shows this in outline; this figure is a simplified version of Fig. 5-1.

In many bacteria the enzymes leading from the carbon source to the common pool of intermediates are inducible. Therefore the cell will not make the enzymes for degrading some particular carbon source unless that compound is available. This control is clearly advantageous to the cell. But there is a further and more subtle control. If *two* carbon sources are available and if one can be used by the bacterium more readily than the other, then the enzymes which degrade the second compound are not formed until the medium is exhausted of the first carbon source. For example, if *E. coli* is presented with a mixture of glucose and lactose, the enzyme lactase will not be formed until the supply of glucose is exhausted. The intermediates in the breakdown of glucose, when their concentrations are sufficiently great, inhibit the synthesis of other enzymes, such as lactase, that would produce the *same* intermediates.

In the case of the biosynthetic enzymes, it can be seen that the cell is faced with the same problem in the converse sense. If one of the building blocks is present in the medium, it is wasteful of protein-synthesizing machinery to continue to synthesize the enzymes for the formation of this building block. Under this condition the cell stops synthesizing these enzymes. But as with the degradative enzymes, a more subtle kind of control is also exercised. In the absence of an external supply of building blocks, the concentrations of these compounds formed by the cell itself regulate the formation of the biosynthetic enzymes. As an example of how this can work, con-

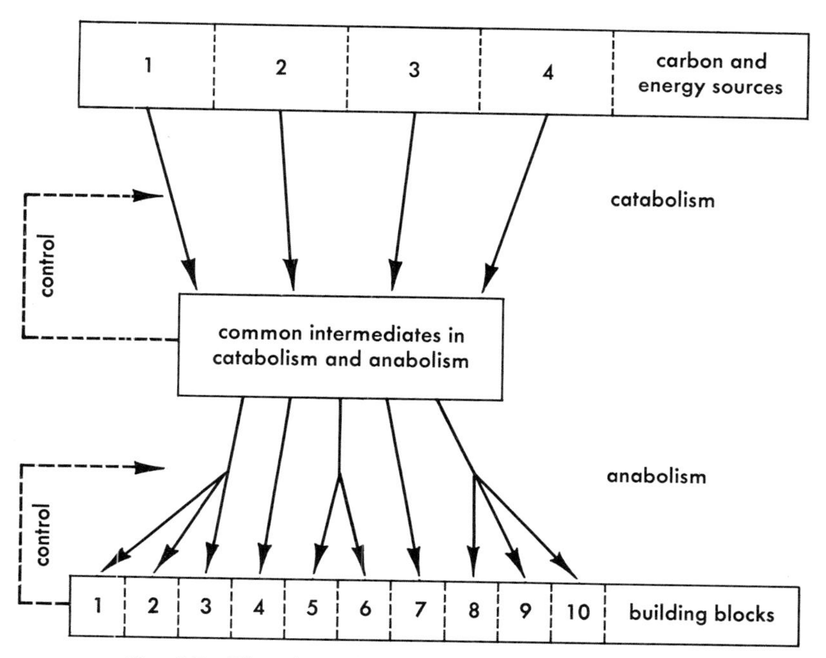

Fig. 6-7. The relationships of catabolism and anabolism.

sider a bacterium growing in a medium with a single source of carbon. The relative amounts of all its enzymes and metabolic intermediates will be mutually controlled, as just described. Now if an amino acid is added to the medium, the growth rate increases. The synthesis of the enzymes involved in the formation of this amino acid will stop; thus, amino acids that would have been used to make these enzymes can be used now to make more of the other enzymes. But if the growth rate increases, the rate at which amino acids are being turned into protein increases and the concentration of the amino acids tends to decrease. The amino acids thus regulate the synthesis of the enzymes that make them; therefore, when their concentration falls, the biosynthetic enzymes increase in amount—thus tending to restore the pool of amino acids. In a short while the cell will be in a new, steady state, growing exponentially at a higher rate.

Induction and repression of enzyme synthesis is a rather sluggish way of regulating cell metabolism. It is as if the temperature of a building were controlled by building or dismantling furnaces. But bacteria have another way of regulating their metabolism, which can be likened to the use of a thermostat in heating a building. In many cases, the final product of a series of enzyme reactions not only represses the synthesis of the enzymes but also inhibits their activities. In this manner, the cell avoids using its supply of carbon and energy producing a building block that is present in the environment.

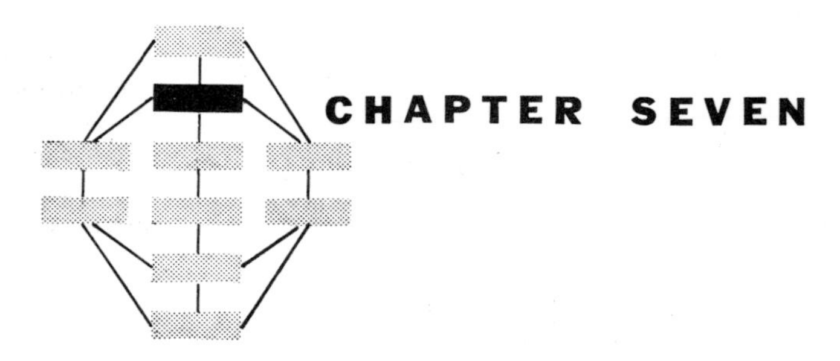

CHAPTER SEVEN

GENETIC SYSTEMS OF THE PROTISTA

Sexual reproduction occurs so uniformly in familiar plants and animals that it is hard not to believe that it is universal and that it needs no explanation. An examination of reproduction in the Protista reveals, first of all, that many organisms have no sexual reproduction, and second, that the kind of sexuality seen in higher plants and animals is only one of a great variety of modes of sexual reproduction. A brief look at some of the patterns of sexual reproduction among the fungi and algae will permit a better understanding of the biological meaning of sex. It will then be possible to inquire into the occurrence and significance of sexual reproduction in the bacteria.

SEXUAL REPRODUCTION IN HIGHER PROTISTS

In all its manifold variations, sexual reproduction has one constant feature: the union of the genetic material of two individuals to form the genetic material of another individual. The two cells that contribute genetic material are the *gametes,* and the immediate product of the fusion of the gametes and their nuclei is the *zygote.* The amount of genetic material in the zygote nucleus is thus twice that in each of the gametes. The zygote will be referred to here as *diploid* and the gametes as *haploid,* although in some cases this is not strictly true. At some time during the life cycle of a sexually reproducing organism, haploid gametes must be formed from the diploid zygote or from the organism that arises from the zygote. All of these facts are covered in *Genetics,* in this series. This generalized life cycle is shown in Fig. 7-1A.

Several variations on this theme are possible and all of them occur among the protists; these variations are outlined in Fig. 7-1. Some specific examples will now be considered.

The water mold *Allomyces* exemplifies the pattern shown in Fig. 7-1B. It was stated in Chapter 2 that the water molds, or aquatic phycomycetes, are the most primitive of the three main groups of fungi. *Allomyces* is a small fungus consisting of a few simple rhizoids imbedded in the substrate, which is usually a bit of decaying plant material floating in the water. Rising up from the rhizoids are slender and slightly branched hyphae. The reproductive organs arise at the tips of the hyphae. As indicated in the figure, the haploid and diploid phases of *Allomyces* are equally well developed and prominent, and apart from the nature of the reproductive structures they bear, are hardly distinguishable. The haploid *gametophyte* produces two morphologically different sorts of gametes: male and female. The gametes are produced in male or female *gametangia*. The gametes are motile, and after being released from the gametangia swim about for a while before coming together in pairs. The male and the female gametes of each pair then fuse to form a single cell; the two gamete nuclei then fuse to form the zygote (diploid) nucleus. Gametes from a single plant can form zygotes. *Allomyces* is thus self-fertile or *homothallic*. The zygote settles down and germinates to produce the diploid *sporophyte*. Both the gametophyte and sporophyte are well-developed multicellular plants; hence, both the haploid and the diploid nuclei can divide mitotically. Both kinds of plants can reproduce themselves by the formation of zoospores, which are haploid in the gametophyte and diploid in the sporophyte. The sporophyte can form another kind of motile reproductive cell that gives rise to a gametophyte. Clearly, meiosis occurs during the formation of these spores; they are, therefore, termed *meiospores*.

The life cycle of the common mold *Rhizopus* (Fig. 7-1C) introduces two new features. In the first place, the diploid phase in *Rhizopus* is greatly reduced and does not have an independent existence as it does in *Allomyces*. Immediately after its formation, the zygote nucleus undergoes meiosis to produce haploid *zygospores*. The zygospores germinate to re-form the gametophyte. In this mold, only haploid nuclei divide mitotically.

The life cycle of *Rhizopus* differs from that of *Allomyces* in another way: the gametes produced on a single plant of *Rhizopus* cannot form a zygote. This condition is called *heterothallism* or self-incompatibility. There are two mating types or sexes of *Rhizopus*. However, since the gametes are morphologically the same, these mating types are termed plus and minus rather than male and female. Two gametes of different mating types will fuse to produce a zygote.

Mating types occur very frequently in fungi that reproduce sexually. In many cases there are more than two mating types, and indeed in some Basidiomycetes there are hundreds. The only requirement for zygote formation in these cases is that gametes be of *different* mating types.

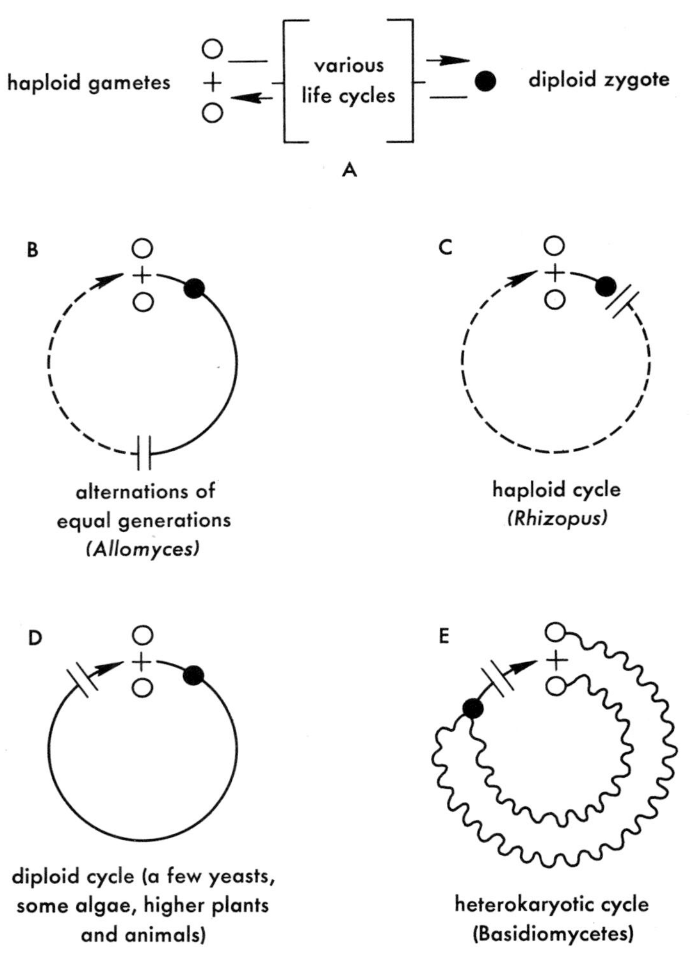

Fig. 7-1. The life cycles of the protists. Gametes are indicated by open circles and zygotes by solid circles; the diploid phase by a solid line and the haploid phase by a dashed line; the position of meiosis is shown by two parallel lines. The individual life cycles are discussed in detail in the text.

The two kinds of life cycle considered so far (the diploid–haploid cycle of *Allomyces* and the haploid cycle of *Rhizopus*) are widely distributed in the fungi and algae. The kind of life cycle to which we are most accustomed (wherein the diploid organism predominates) is somewhat more restricted in the Protista. Some few yeasts (ascomycetes) have a true diploid (Fig. 7-ID) cycle, in which the reduction division and formation of ascospores follows rather than precedes the vegetative growth of the plant. The rest of the Ascomycetes have the haploid cycle. In the Basidiomycetes there occurs a life cycle that is, in many ways, the same as the diploid cycle (see Fig. 7-1E). The key feature here is the formation of heterokaryons, which can be looked upon as the product of a *cellular* fusion of gametes that is not immediately followed by *nuclear* fusion. The two kinds of nuclei continue to divide mitotically, producing a plant that has all the genetic material of both parents, still in separate bundles. Nuclear fusion and the formation of the zygote eventually occur. The zygote immediately undergoes meiosis to produce the haploid *basidiospore*. The related cytological features of the basiomycetes have already been discussed in Chapter 2.

In the algae the diploid cycle (Fig. 7-1D) is more common than it is in the fungi. Several green algae and one order of the brown algae have this kind of life cycle. Even more widespread in the algae is the *Allomyces* type of cycle, with an alternation of gametophytic and sporophytic generations. In some algae, the gametophyte and sporophyte are morphologically indistinguishable and equally prominent; in others, one or the other may be somewhat reduced.

Before turning to sexual reproduction in the bacteria, we should try to see what sense we can make out of these different life cycles. Broadly speaking, it can be said that a life cycle in which the haploid phase predominates is restricted to simple organisms that do not go through an elaborate program of differentiation. Rarely does the diploid cycle occur in simple organisms—the most conspicuous example being the diploid yeast. On the whole, however, diploidy is the rule in larger and more complex organisms, because the orderly growth of a complex organism requires that a very precise time sequence of physiological and morphological changes occurs. This precise developmental sequence can only come about in an organism whose cells will remain genetically stable over many cell divisions. As stated previously, genetic material is inherently mutable. In a haploid organism any mutation is immediately expressed; in a diploid organism mutations are not expressed (unless they are rare dominant mutations) until the same two mutations come together in the one organism. Thus, a haploid genome cannot provide the stability necessary for carrying out a complex developmental sequence, whereas a diploid genome can provide this stability.

Diploidy gives stability; but, by the same token, it reduces variability, which is the raw material of evolution. Sexual reproduction assures that the variability stored in the diploid parent as a recessive mutation is occasionally allowed to be expressed. This analysis leads to the conclusion that diploidy is necessary for the success of large and complex organisms, and that sexual reproduction is then required to permit variation to occur.

The haploid state is not without its advantages: a haploid organism is able to respond very rapidly by mutation and selection to a changing environment. This is especially true in organisms with a very high growth rate and this is precisely the kind of organism in which haploidy predominates: bacteria, simple fungi and algae.

SEXUAL REPRODUCTION IN BACTERIA

In all fundamental aspects the genetics of bacteria is the same as that of other organisms. Each heritable property or character of a bacterium is determined by a particular portion of its genetic material which, as in other organisms, is deoxyribonucleic acid (DNA). As far as is known, bacteria are normally haploid. The genetic material of a bacterium can mutate or change and since bacteria are haploid, the mutations are immediately expressed. As we shall see presently, sexual recombination occurs in some bacteria and it is possible to construct a genetic map of the bacterial chromosome; as in other organisms the map is linear.

In what follows it is assumed that the student is familiar with basic ideas of the genetics of sexual reproduction; only those features peculiar to the bacteria are stressed. Until quite recently sexual reproduction was not known to occur in the bacteria; at the present time three different processes are known to occur in bacteria by which new combinations of genetic characters can arise. These three processes are: conjugation, transformation and transduction.

Conjugation

Because it is most closely related to sexual reproduction in other organisms, *conjugation* will be considered in most detail. It is distinguished from transduction and transformation by the fact that the parental cells must come together in pairs (conjugate). Conjugation occurs in several different bacteria but is best known in *Escherichia coli*. The genetic material of *E. coli* is arranged in a single chromosome or linkage group. As in other organisms, the parts of the genetic material with which the properties of the bacteria are associated (the genes) are arranged in a linear order.

There are two mating types or sexes in *E. coli*. Both sexes are capable of growing indefinitely by binary fission; sexual reproduction is certainly not obligatory. When a suspension of male cells is mixed with one of female cells, the two kinds of cells come together in pairs or conjugate. The male then injects its chromosome into the female, and subsequently dies. The transfer always begins at one end of the chromosome and proceeds in order. In Fig. 7-2, for example, the end marked O enters first, followed by the regions marked *a, b, c,* etc., in that order. The transfer can be interrupted at any time by separating the partners of the conjugating pairs by violent shaking of the suspension. In this case the only properties of the male which can appear in the progeny are those which have been transferred before the interruption. Interruption of the transfer occurs sooner or later in the great majority of conjugating pairs, even when they are not artificially separated.

In most cases, therefore, the zygote contains a complete female chromosome but only a partial male chromosome. The new combinations of genetic material occur presumably by the same process of crossing over as in other organisms, although in bacteria there is no cytological evidence for this. Indeed, there is no evidence for meiosis in bacteria; but, in a sense, the complex mechanism of meiosis is really not needed here because the cells have only one chromosome. The entire sequence is outlined in Fig. 7-2.

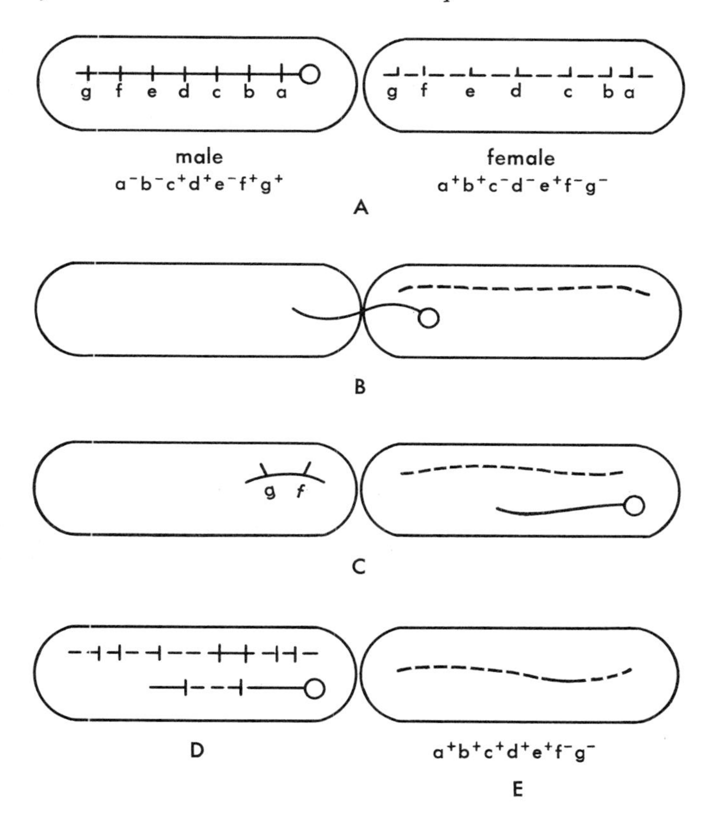

Fig. 7-2. Conjugation and zygote formation in bacteria. A: Two conjugating cells are shown diagrammatically. The letters refer to genetic characters in which the two cells differ. B: The male cell has begun to inject its genetic material into the female, beginning with the end marked O. C: The injection has stopped before all of the male's genetic material has been transmitted. This is the usual case. The female cell is now a partial zygote. D: An exchange of genetic characters between the two sets of material in the zygote. E: The cells formed after duplication of the *complete* genetic material in the zygote. This cell resembles its female parent in characters *a, b, e, f,* and *g;* and its male parent in characters *c* and *d.* Other kinds of recombinant progeny can be produced, depending upon the position of the exchange genetic material.

Certain characters of *E. coli* do not follow this pattern. Most important is the character of sex itself. In a mating such as described above, none of the progeny would be female. Maleness is determined by the presence of extrachromosomal genetic particles; to be female, a cell must contain none of these particles. The particles are very rapidly transferred from male to female cells during conjugation. In fact they are transferred much faster than is the chromosome. Thus, if the cells are separated very soon after conjugation, all will be male, even though very little chromosomal genetic material is transferred. Although cells which contain the male-determining particles are males in the sense that they will not conjugate with other males, they are unable to transfer their chromosomes to females. In other words, they are sterile males. These males become fertile if a male-determining particle attaches itself to the chromosome. This attachment occurs rarely, but it is relatively stable; hence, it is possible to obtain populations of fertile males without much difficulty. It is these fertile males that take part in the conjugation discussed earlier. The end of the chromosome to which the male-determining particle is attached is the last to be transferred. Because transfer is in most cases only partial, this end is only rarely transferred. The non-appearance of fertile males among the progeny is thus explained. As already mentioned, the progeny are either female or, if the male-determining particle has been transferred, males.

The peculiar behavior of the male-determining particle, especially the fact that it can exist alternately on the chromosome or in the cytoplasm, is similar in many ways to that of certain bacterial viruses. More will be said about this kind of genetic particle, or *episome,* in the discussion of bacterial viruses in Chapter 8.

Transformation and Transduction

These differ from conjugation in the mechanics of the transfer of genetic material and in the limited amount of material that is transferred. In *transformation,* the DNA of one strain of bacterium is freed from the cells and directly enters another cell; the DNA moves from cell to cell in solution. Transformation is a laboratory procedure and may not occur in nature. In the laboratory the DNA of one strain is extracted and chemically purified. Then the DNA is mixed with cells of another strain, differing from the first in one or more heritable properties. These cells will continue to grow and divide. Among their progeny there will be a few cells with properties of the strain from which the DNA was extracted. Only rarely will a single cell be transformed for more than one property. When this does happen it is because those genes of the donor cells that determine the transformed characters are very close together.

In *transduction,* genetic material is transferred from one cell to another by means of a temperate bacteriophage. (The synthesis of bacteriophage is discussed in Chapter 8.) Temperate bacteriophage are first produced in one strain of a bacterium of known genetic constitution; for example, a strain that may be unable to make the amino acid histidine. The bacteriophage are allowed to infect a different strain; for example, one that is able to synthesize histidine. Among the bacteria surviving the infection will be found a few that have acquired one of the properties of the first strain. In our example, such cells would be unable to make histidine. Only infrequently will more than one property be transferred. Apparently what happens is that during the synthesis of the bacteriophage, small pieces of the genetic material of the bacterium become incorporated into the bacteriophage. When the phage infects another cell, the piece of genetic material from the first bacterium becomes integrated with that of the second. If this bacterium survives the infection, its progeny will acquire the property determined by this bit of genetic material. As in transformation, only small pieces are transferred; hence it is very improbable that a single cell will be modified in more than one property.

Gene Transfer and the Natural History of Bacteria

One or more of the three mechanisms of gene transfer has been found in numerous bacteria, and there is no reason to suppose that they are not present in all the groups of bacteria. What then is their importance to bacteria in nature?

At the moment, there is little direct evidence of these processes occurring in nature. They may be merely laboratory curiosities, which reflect certain properties inherent in genetic material. Although this may be true of transformation and transduction, it seems improbable that so complicated a process as conjugation is without biological significance.

Assuming that gene transfer and recombination occur in nature, of what importance are they? Sexual reproduction is an important source of variation in diploid organisms, but in haploid organisms, such as bacteria, variation is assured by mutation. If only the amount of variation were important, recombination would not be an advantage to haploid organisms. Mutation, however, does not permit *groups* of genes to be tried out in new environments and in new combinations of other characters. Conjugation permits recombination among relatively large units. In this connection it is significant that in many bacteria, genes which control formation of a group of enzymes catalyzing a sequence of reactions are themselves linked in the chromosome. In conjugation the whole group can be transferred together.

So far variability of bacteria has been discussed; equally important is their stability. By stability is meant the continued association of various characteristics. It is possible to isolate from nature the same bacterium time after time; for example, almost invariably, strains of *E. coli* isolated from nature can ferment lactose. But it was seen in Chapter 6 that a single mutation can result in the loss of ability to ferment lactose. How is this property which seems so unstable genetically, retained in nature? The principal reason for stability of bacterial types is the intense selection to which they are subject. Because of the very rapid growth of bacteria, selection operates rapidly and ruthlessly in a bacterial population. A given type of bacterium is adapted to a particular environment; any mutation leading to a less successful type is quickly eliminated by selection. It follows then that in order to have a real understanding of a bacterium and to be able to attempt to classify it, one must possess a knowledge of its particular environment.

Another phenomenon that contributes to the genetic stability of bacteria depends upon the fact that in any population of bacteria, mutants can arise which are slightly better adapted—that is, they grow a little faster—than the original population. These better adapted mutants will, if the culture is allowed to grow exponentially for a long time, displace the original population. An even better adapted mutant can now arise in this second population and displace the first mutant. Now what will be the effect of such *periodic selection* of better and better adapted mutants on other kinds of mutants? Since both kinds of mutation are rare, it is unlikely that they will occur together in the same cell. In all probability then, a better adapted mutant will displace any other kind of mutant. This can be made more clear, perhaps,

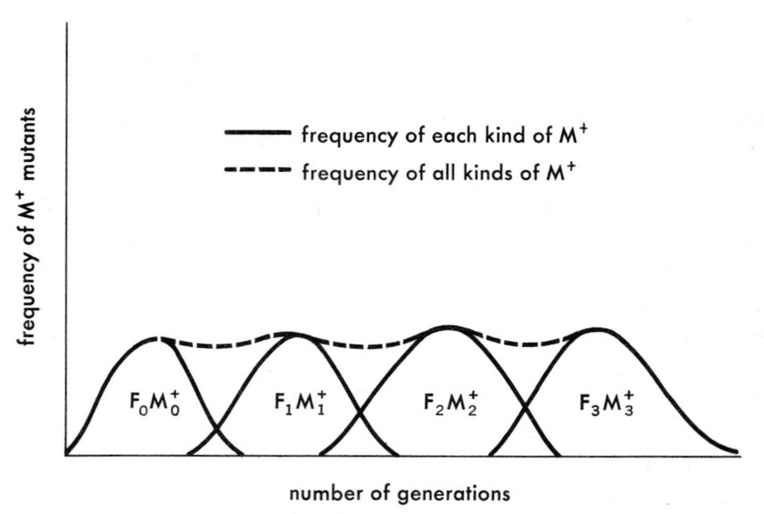

Fig. 7-3. Periodic selection. See text for details.

by using symbols. The growth rate mutants will be called F_0 (the original population) F_1, F_2, F_3. The second kind of mutants will be called M^+. The original population then is symbolized by F_0M^- and the sequence of events is:

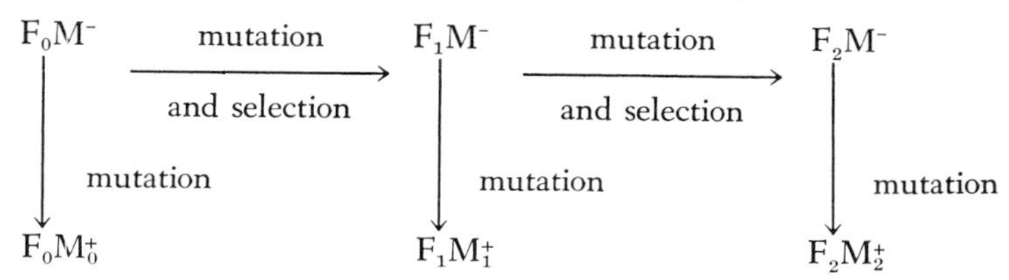

The different kinds of M^+ mutants (F_0M^+, F_1M^+, etc.) will rise and fall periodically (see Fig. 7-3). But the whole class of M^+ mutants will remain fairly constant in number.

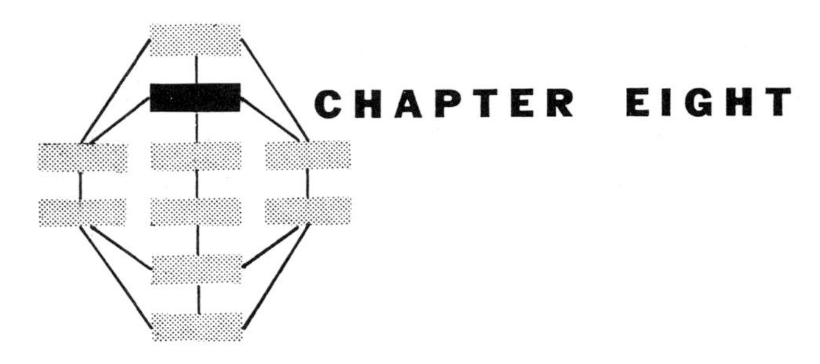

BACTERIOPHAGE AND VIRUS

—The hen is the egg's way of making another egg

—Samuel Butler

Although Mr. Butler's view of how the egg looks upon the hen is extreme, it is not so extreme as the virus' view of its host. We are concerned here with the point of view of the virus. Viruses are self-replicating units that, unlike other organisms, contain only one kind of nucleic acid—either ribonucleic acid or deoxyribonucleic acid. They are obligate intracellular parasites that infect bacteria and the cells of plants and animals. As a rule, any one virus can infect only one kind of host; that is, a virus is specific for its host.

All viruses are extremely small, but there is a considerable range of sizes among them. One of the largest viruses is that of cowpox, which is used in vaccination against smallpox. It is a cube about 20 $m\mu$ on a side. A small virus may be no more than 2 $m\mu$ in diameter. The sizes of a few well-known plant and animal viruses are given in Table 8-1. Viruses have very simple morphologies, since most viruses are composed not of millions of molecules of hundreds of different kinds but of only a few molecules of only several kinds. As will be shown, these few molecules are able to infect a cell and cause it to produce hundreds of new virus particles. Their minuteness and infectivity are combined in the usual (and not wholly satisfactory) definition of viruses as "submicroscopic infectious particles." One trouble with this definition is that it ignores the role of the host. The emphasis here will be on the viruses that infect bacteria—the bacteriophages. Because the host cells of these viruses are so easy to grow in large numbers under carefully controlled conditions, more is known about the biology of bacteriophages than of plant or animal viruses.

Bacteriophages are not rare; probably there is a phage that can grow on any bacterium. Only a few have been studied extensively, but so far as is known all bacteriophages have the same general characteristics. The most extensively studied bacteriophages attack *Escherichia coli*, a common bacterium usually present in large numbers in raw sewage. These "coli phages" are easy to isolate. A culture of *E. coli* is inoculated with a small amount of sewage that has been filtered to remove bacteria. The phage particles are so small they will pass through the filter. The culture will continue to grow but if "coli phages" are present in the filtrate, the culture will eventually lyse; that is, it will become clear. A small amount of this lysed culture can be filtered to remove any remaining bacteria and will cause the lysis of another culture. This process can be repeated an indefinite number of times. Thus, the ability to lyse is not lost even after many transfers involving a very great dilution of the original filtrate. Therefore, it can be concluded that the lytic agent multiplies. Because the agent can pass a filter that retains bacteria and because microscopically visible particles are absent, it is apparent that the bacteriophage is very small. Phage particles cannot be seen in the ordinary microscope, but their ability to lyse bacteria provides a means of counting them. This can be done by mixing a suspension of sensitive cells and phage particles and spreading this over the surface of a solid medium. The growth of the bacteria will produce a turbid film of growth peppered with clear areas or plaques. The plaques represent the lysis caused by the multiplication of a single phage particle.

With the advent of the electron microscope it became possible to examine the morphology of bacteriophages. The "coli phages" are composed of two parts: a head and, attached to it, a short tail. The head is a hexagonal prism with pyrimidal ends; the tail, which is about as long as the head but considerably narrower, is hexagonal in cross section (Fig. 8-1). Chemically the phage is composed *only* of deoxyribonucleic acid and protein. The DNA is within the head; the wall of the head and the tail are protein. Each kind of phage has its own particular composition; one of the most studied coli phages contains about 6×10^{-13} μg of DNA and an approximately equal amount of protein. The DNA is probably a single molecule; there may be several tens of protein molecules.

LIFE HISTORY OF LYTIC PHAGES

Let us now examine in more detail the development of phage in an infected bacterium. The process can be divided into three stages: (1) infection, (2) growth, and (3) lysis. Only one phage particle is necessary for infection; however, infection can be by more than one particle. The infection of the host cell occurs in two stages: attachment of the phage and the infection itself. The tip of the tail (which is really a proboscis) attaches to the cell wall of the host. The phage can stick to the wall only at certain places

TABLE 8-1

Properties of Some Better-Known Plant and Animal Viruses

	Particle size[a]	Dimensions	Shape	Composition (percent) Nucleic acid	Protein	Other
tobacco mosaic	40×10^6	180A × 3000A	needles	RNA 66	94	none
tomato bushy stunt	9×10^6	300A	spheres	RNA 16	80	none
turnip yellow mosaic	5×10^6	220A	spheres	RNA 35	65	none
poliomyelitis	10×10^6	270A	spheres	RNA 25	75	none
rabbit papilloma	50×10^6	450A	spheres	DNA 9	90	none
influenza	280×10^6	800A	spheres	RNA 1	70	carbohydrate 23 lipid 5
cowpox (smallpox vaccine)	ca 3×10^9	280 × 220 × 220A	brick-shaped	DNA 6	90	lipid 6

[a] The weight of a single virus particle relative to the weight of a hydrogen atom.

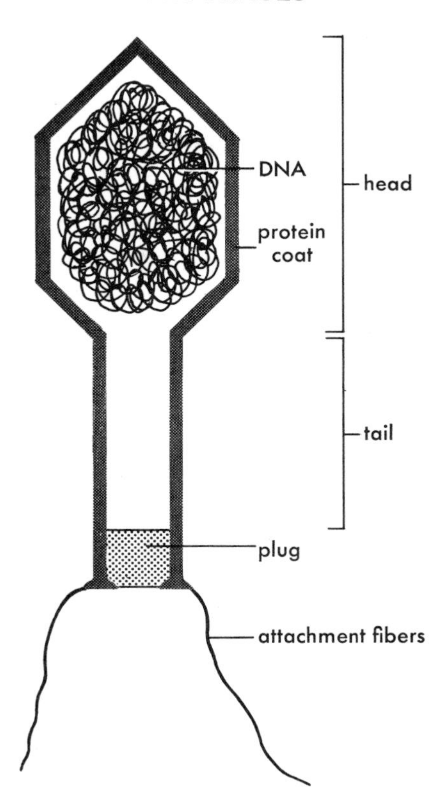

Fig. 8-1. A very diagrammatic representation of a "longitudinal section" of a bacteriophage particle. The attachment fibers anchor the phage to the cell wall of the host; the tip of the tail contains an enzyme that attacks the rigid layer of the host's wall; the main portion of the tail then apparently contracts, injecting the DNA into the cytoplasm of the bacterium.

called receptor sites; different phages have different specific receptor sites. Bacteria can become resistant (by mutation) to infection by a particular phage. When this happens the phage no longer is able to attach to the cell wall. Apparently the configuration of the receptor has changed so that the phage no longer "fits." Different kinds of phage can attach to one host, but they do so by attaching at different sites.

After the phage attaches to the cell wall the actual infection occurs. First a small area of the cell wall is dissolved by a phage enzyme. Then the DNA of the phage is injected into the host cell. (The mechanism is still obscure.) Only DNA is injected; none of the protein of the phage coat enters the cell. Since the infected cell will go on to form 100 or more complete phage particles, it is evident that DNA alone is sufficient to cause not only its own replication but also the formation of the protein part of the phage particles. The way in which protein forms is certainly not clear; however, it is known that immediately after injection of the DNA, the metabolism of the host cell is profoundly altered. The synthesis of host-specific macromolecules ceases; that is, the bacterium stops forming those molecules that made it a cell of a particular kind. After a time, phage-specific macromolecules are formed—DNA and protein. The proteins formed include not only the structural proteins of phage, but also certain enzymes necessary for the building blocks (purines) peculiar to the phage DNA. These proteins (enzymes) are the first phage-specific molecules to be made. Thus, the energy metabolism and biosynthetic mechanisms of the host have been removed from the control of the bacterial genetic system and are now controlled by the phage genetic system. The synthesis of phage protein and phage DNA go on independently. After a few minutes, complete phage particles begin to be assembled from the already-formed DNA and protein; when the number of phage particles reaches about 100 to 200, the cell bursts and the phage are released. This is lysis. The entire process—infection to lysis—takes about 20 minutes in *E. coli.*

The phage particles, it should be noted, do not divide or multiply; one phage particle does not give rise to two and these in turn to four. Rather, the complex "phage + host cell" produces hundreds of phage particles. The self-replicating unit is neither the host cell nor the phage; rather it is the complex of phage genetic material and host cell. Since this complex is an unusual organism, it may be referred to as a *metaorganism.* In ordinary cell growth, the genetic system of the cell determines the specificity of the metabolic activities of the cell. It thus determines the specificity of what is replicated. In the "phage + host cell" metaorganism, the phage is the genetic system and determines the specificity of the replication; but it can operate only by making use of the machinery of the host cell.

In many respects the DNA of phage is a perfectly ordinary genetic system. It can change or mutate as does other genetic material. Also, two distinct phage genetic systems can, during infection of a common host cell, recombine to form new genetic systems that are different from either. The genetics of phage is covered in more detail in *Genetics,* in this series.

PLANT AND ANIMAL VIRUSES

The biology of plant and animal viruses is basically the same as that of a typical bacteriophage. Plant and animal viruses differ from bacteriophage and from one another in chemical composition, shape, and the kind of host cell attacked.

A few properties of some of the better-known plant and animal viruses are shown in Table 8-1. Probably the most intensively studied of these is tobacco mosaic virus (TMV for short). As with other viruses, this one gets its name from the disease it causes. As can be seen in the table, TMV contains RNA and protein only. A virus contains only one kind of nucleic acid, either DNA or RNA. All viruses also contain protein, and many have no other component. Some of the larger viruses contain, in addition, lipid or carbohydrate. TMV is a long, thin, rod-shaped particle. The particle is a hollow tube of protein with the RNA inside. The protein tube or coat is formed of identical protein molecules in a tightly wound helix. Like the DNA of a bacteriophage, the RNA of TMV contains all the genetic information necessary for the formation of a complete virus particle. It is possible to remove the protein coat from the virus and separate it from the RNA. The naked RNA is still capable of infecting tobacco plant cells; and the infection results in the production of complete virus particles. The naked RNA is, however, much less effective than the complete virus. The protein coat apparently protects the RNA and somehow facilitates its entry into the host cell. Recently, TMV has contributed greatly to our understanding of the functioning of genetic material; these aspects are considered in *Genetics*, in this series.

Many viruses, like bacteriophage, are crystallizable. This ability to form crystals means that the particles of a particular virus have a uniform and orderly structure.

LYSOGENY

The kind of phage just discussed is called lytic or virulent phage. The temperate phage is another kind of bacteriophage that, in contrast to a lytic phage, does not invariably lyse its host. If a suspension of host cells is mixed with a suspension of a temperate phage, only a fraction of the cells will lyse; the remainder continue to grow and are apparently normal. The development of the phage in the cells that lyse is exactly as described previously. The cells that remain unlysed, however, have two peculiar properties: first, they are completely resistant to infection by the same phage; and second, these cells can induce lysis of cells that are sensitive to the phage. Such cells are termed lysogenic.

Careful examination of a culture of a lysogenic bacterium reveals that occasionally (once in about every thousand divisions) a cell will lyse and release free phage particles. Any cell of a lysogenic culture may suddenly lyse; however, when lysogenic bacteria are artificially broken open, no phage particles can be found in them. Thus, lysogenic bacteria are not full of mature phage particles; instead, they have the ability to form phage. This ability is transmitted to all of their progeny.

This property or ability is closely associated with the genetic system of the bacterium; indeed, lysogeny (the inherited ability to produce phage) is genetically no different from any other property of the bacterium. (See *Genetics* in this series for a more detailed discussion.)

Temperate bacteriophage can be in either of two alternative states: (1) they may be free phage particles, which can infect and lyse a bacterium or (2) they may be an integral part of the genetic material of a lysogenic bacterium. The latter state is termed the "prophage"; both have a common vegetative state during which phage particles are formed. In ordinary circumstances, a lysogenic bacterium containing prophage only rarely produces phage particles. In certain circumstances—for example, after irradiation with ultraviolet light—all the cells of a culture of a lysogenic bacterium form free phage. This process is called "induction of the prophage." The development of free phage in an induced lysogenic bacterium closely resembles that of lytic phage, except, of course, infection is unnecessary.

As stated earlier, lysogenic bacteria are immune to infection by temperate phage. This immunity is specific: a bacterium lysogenic to one kind of phage may be infected with another kind; the second temperate phage may either lyse or lysogenize the cell. Thus, a single cell may carry more than one prophage. Immunity is not the same as resistance to a lytic phage. It has been shown that a cell is resistant because of the inability of the phage to attach to the cell wall. On the other hand, in the case of an immune cell the phage does attach to the wall and even injects its DNA into the cell, but the immune cell does not produce new phage; the development of the vegetative phage is repressed. This repression of phage development is known to be a property of the cytoplasm of the host. The cytoplasm of a cell carrying a prophage is so constituted that the development of the vegetative phage is inhibited. In other words, that portion of the bacterial genome called "prophage" is inactive. A temperate phage can mutate so as to be insensitive to this inhibition. Such a mutant always lyses the host cell and is said to be virulent. Genetic elements that can exist in two alternative states are known as *episomes*. It was pointed out in Chapter 7 that the male-determining factors in *Escherichia coli* are another example of episomes.

If you will now reconsider what was said in Chapter 6 concerning the formation of induced enzymes, you will remark a most striking and satisfying uniformity: in the formation of both proteins (induced enzymes) and larger structures (bacteriophage particles) there seems to be a fundamentally similar mechanism. In both cases the expression of the genotype can be repressed by substances within the cytoplasm. Both cases give examples of nonrepressed forms: constitutive enzymes and virulent strains of temperate bacteriophage. The discovery of these parallels in two such seemingly dissimilar phenomena is one of the most impressive advances of modern biology. In this case, repression is made manifest by the absence of extrachromosomal male particles in *fertile* male cells.

INDEX